W9-AUU-948

NATIVE BROMELIADS OF FLORIDA

Date: 6/24/20

584.85 LUT
Luther, Harry E.,
Native bromeliads of Florida

PALM BEACH COUNTY
LIBRARY SYSTEM
3650 SUMMIT BLVD.
WEST PALM BEACH, FL 33406

Guzmania monostachia, Fakahatchee Strand Preserve State Park (K. Marks photo)

NATIVE BROMELIADS OF FLORIDA

Harry E. Luther
David H. Benzing

in association with
Marie Selby Botanical Gardens

Pineapple Press, Inc.
Sarasota, Florida

Copyright © 2009, 2016 by Marie Selby Botanical Gardens

All rights reserved. No part of this book may be reproduced in any form or by any means, electronic or mechanical, including photocopying, recording, or by any information storage and retrieval system, without permission in writing from the publisher.

Inquiries should be addressed to:

Pineapple Press, Inc.
P.O. Box 3889
Sarasota, Florida 34230

www.pineapplepress.com

Library of Congress Cataloging-in-Publication Data

Luther, Harry E., 1952-
Native bromeliads of Florida / Harry E. Luther and David H. Benzing. -- 1st ed.
p. cm.
Includes bibliographical references and index.
ISBN 978-1-56164-967-9 (pb : alk. paper)
1. Bromeliaceae--Florida. 2. Endemic plants--Florida. I. Benzing, David H. II. Title.
QK495.B76L88 2009
584'.8509759--dc22
2008046496

First Edition
10 9 8 7 6 5 4 3 2 1

Design by Shé Hicks
Printed in the United States of America

ACKNOWLEDGMENTS

The authors thank Bruce Holst, Valerie Renard, Karen Norton, Joanne Miller, Jay Staton, Ken Marks, Linda Grashoff, and Phil Nelson.

Tillandsia setacea, Manatee River (B. Holst photo)

Color Photographs (CPs) between pages 64 and 65

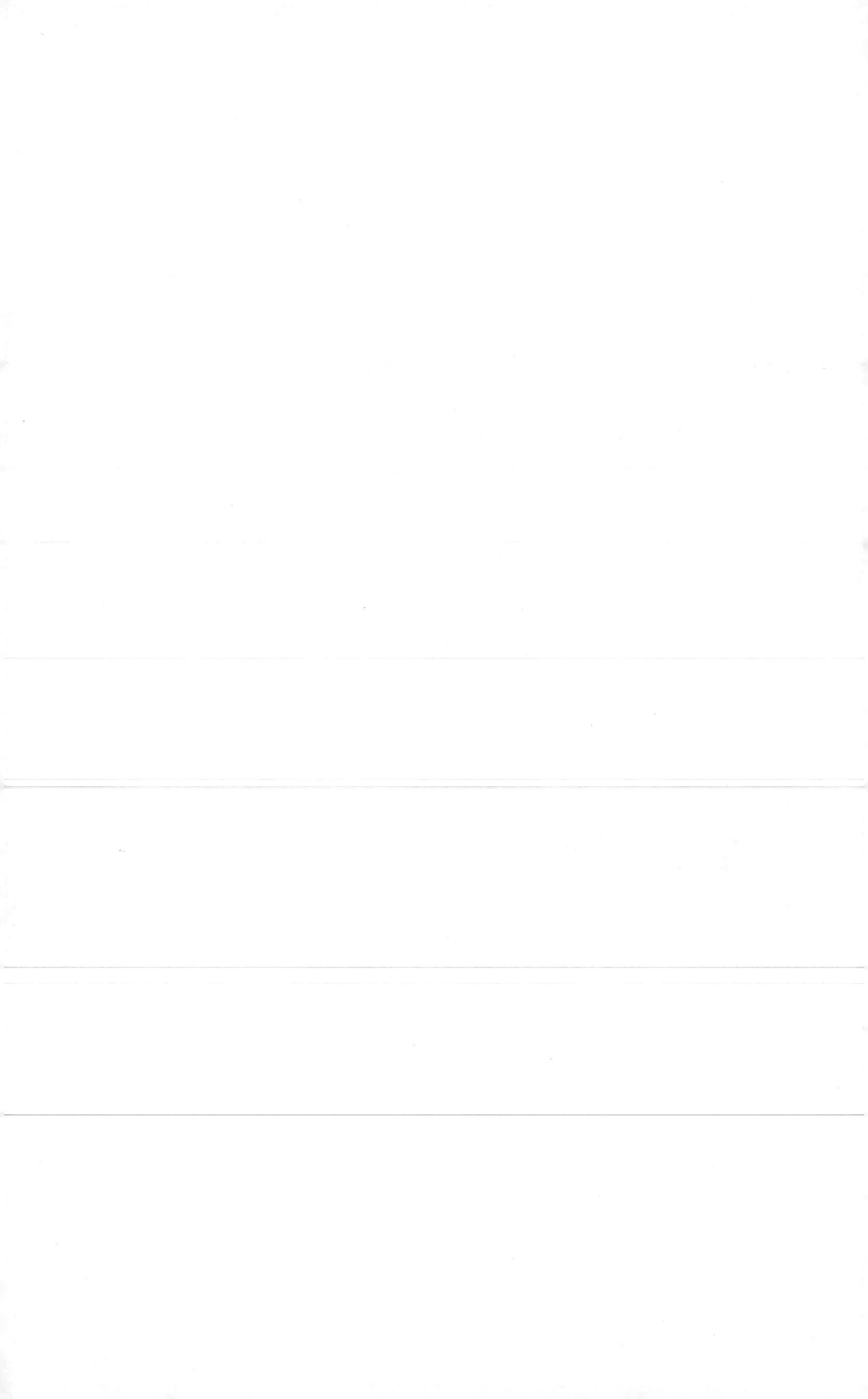

Guzmania monostachia, Fakahatchee Strand Preserve State Park (J. Staton photo)

INTRODUCTION

This book provides the means to identify Florida's native bromeliads and appreciate their many unconventional characteristics. Though there are more than 3400 species of bromeliads, Florida can only boast sixteen of them, plus two natural hybrids. Only one of these sixteen natives is endemic, that is, found only in Florida.

Florida's vegetation consists of a large number of species and is exceptionally varied due to the state's diverse climates, geology, and hydrology—not to mention its repeated colonization by waves of botanical immigrants arriving from the north and the Caribbean. Throughout the state, soil chemistry ranges from acid to alkaline and from fertile to impoverished. Moisture supplies also run the gamut from being abundant year-round to effectively absent through months-long winter dry seasons. While frost is an annual event in the north, it virtually never occurs in the southernmost Keys.

Among the many kinds of plants that reside in the state's numerous types of habitats are the bromeliads. These so-called "air plants" thrive on trees and shrubs as epiphytes, which means they have no soil-penetrating roots, or in some cases, virtually no roots at all. Unlike Florida's three mistletoes, none of them is a parasite, as they use their woody hosts only for mechanical support.

Florida's bromeliads occupy ranges largely defined by winter temperature. The cold-hardiest among them extend northward into Alabama and Georgia, and in the case of Spanish moss, all the way up the Atlantic coast to Virginia. The most sensitive species maintain only the barest of toeholds in the extreme south. The so-called "wild pine" bromeliads fall at various points farther down the hardiness scale, being more vulnerable than Spanish moss, but tough enough to extend most of the way up the state. In essence, they require winter protection, which, within the continental U.S., only peninsular Florida reliably provides. Several of the larger-bodied wild pines lend an unmistakably south Florida character to the cypress swamp forests where they occur so abundantly. Most tenuous of all is *Catopsis nutans,* a species that ranges through Mesoamerica well into northern South America. Its entire presence in the United States amounts to a handful of small colonies in Collier County.

Humans, more than frost, pose the greatest threat to Florida's bromeliads. Except for Spanish moss and closely related ball moss, all of them occur today in numbers much diminished by collectors, land development, wetland drainage, and fire. Should these practices continue, an important component of the state's botanical heritage—along with the orchids, ferns, and the other epiphytes that share the same aerial habitats—will disappear.

To help with identification, a dichotomous key is provided for differentiating the species and natural hybrids. There are also descriptions and images of whole plants and their diagnostic parts; range maps; a glossary of technical terms; and brief discussions of bromeliad anatomy, physiology, and ecology. Also included are instructions on how to cultivate the more ornamental of the state's natives.

Additional references are provided for readers who wish to learn more about these fascinating plants. Bromeliaceae not only ranks among the most biologically novel of the more than four hundred plant families, but is also one of the most thoroughly studied scientifically and among the most highly prized by horticulturalists. Finally, to know about bromeliads is to appreciate substantial facets of Florida natural history.

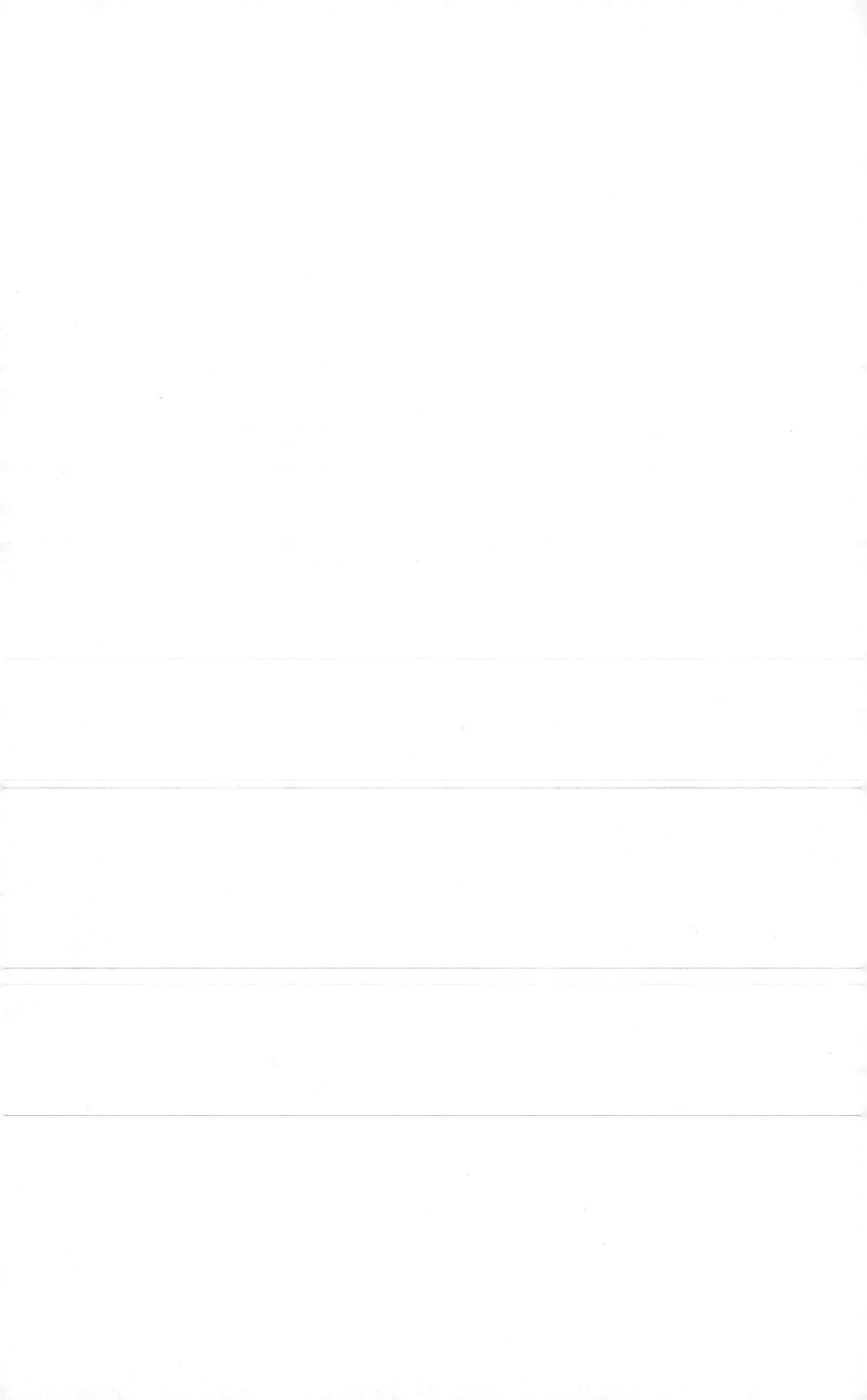

Tillandsia fasciculata, Everglades National Park (B. Holst photo)

ABOUT FLORIDA'S BROMELIADS

Tillandsia setacea (on top of vine) and *T. usneoides* (hanging from vine), Deer Prairie Creek Preserve (B. Holst photo)

WHAT IS A BROMELIAD?

The bromeliads owe their common name to their membership in the plant family Bromeliaceae. Like the names of so many of the other botanical families, this one commemorates a person, and does so in Latin, the traditional language of science. The individual honored in this instance is Olaus Bromelius, a late fifteenth- to early sixteenth-century Swedish physician, who, like the other members of his profession at the time, studied botany as part of his formal training. Doctor Bromelius, as it turned out, developed a more enduring fondness for plants than medicine. Or so it seems given his botanical enshrinement.

With about 3400 species named so far, Bromeliaceae ranks by size somewhere in the middle of the 400 families recognized by taxonomists to accommodate the approximately 350,000 different kinds or species of flowering plants. More noteworthy than this statistic, however, is the family's nearly exclusive tropical American distribution. Just one of its member species occurs beyond the New World, and the home of this solitary exception amounts to a modest-sized wedge of real estate on the other side of the Atlantic Ocean. Its location on Africa's west coast suggests recent colonization from much more bromeliad-rich eastern Brazil, or perhaps from one of the Caribbean islands.

Species that make up a plant family, or any of the animal families for that matter, are assigned to genera (the plural of genus), with each genus containing at least one of these species. Bromeliaceae includes approximately 60 genera, the largest of which contain hundreds of species, and the smallest just a few (or even just one). Until recently botanists studying bromeliad structure, particularly the anatomical details of their flowers and fruits, were convinced that this family consisted of three major subgroups, each of which warranted ranking as a subfamily (Smith and Downs, 1974, 1978, 1979).

However, analyses of DNA structure beginning during the 1980s changed perceptions about bromeliad relationships by indicating that the evolutionary history of Bromeliaceae is more complicated than previously thought (Crayn et al., 2004; Givnish et al., 2007). It now seems that the family is comprised of more than three major subgroups and that some of its species need to be reassigned to different genera. Errors of this nature often take years to correct following their discovery, but someday bromeliad taxonomy will satisfy the general rule that all organisms be classified according to their genetic (phylogenetic) relationships.

All of Florida's sixteen bromeliad species and two named natural hybrids belong to subfamily Tillandsioideae, and within this group to just three of its more than a dozen genera. Genus *Tillandsia* accounts for all but four of the natives and both of the hybrids.

Only two of the Florida bromeliads, both of them tillandsias, vary enough to warrant subdivision of their members into named subspecies or varieties. No members of subfamilies Bromelioideae or Pitcairnioideae naturally reside in Florida today, although they might have in the past for reasons discussed below.

Bromeliaceae, and especially many of the species assigned to subfamily Tillandsioideae, ranks high for horticultural importance (Plever and Brehm, 2003). For absolute dollar value, even the most ornamental of the cultivated bromeliads pales compared to the pineapple, or more specifically, to its fruit. As for other commercial uses for bromeliads, the foliage of a couple of species and several varieties of pineapple yield fiber that supports minor craft industries scattered through rural Latin America. And an enzyme extracted from pineapple fruits that breaks down protein is used to treat burns.

Bromeliad leaves often exceed flowers for adding ornamental appeal. Particularly striking are the species with foliage that features alternating, horizontal bands of maroon and green, or species that display intricate mosaics of tissue visually differentiated by high and lower concentrations of chlorophyll. While flowering, most or all of the shoots of many more species, including several of the Florida natives, color up to bright orange, red, or pink. Dense layers of light-reflecting hairs grant still other bromeliads their characteristic silvery gray appearance (**CP 1**).

Leaf thickness is often considerable, but not enough to designate any but the most drought-hardy of the terrestrial bromeliads as bona fide succulents. Shape varies as well, especially between those bromeliads that do and those that do not impound moisture in leafy tanks. Species with narrow, tapering, more or less upright foliage that forms vase-shaped shoots constitute the so-called "wild pines"; the designation "air plant" tends to be reserved for the bromeliads equipped with softer leaves and no impoundments (**CPs 1, 2**).

Plant physiologists appreciate the bromeliads for how the toughest among them tolerate daunting growing conditions. Especially impressive are the dry-land epiphytes and their relatives that root on bare rock, the so-called "lithophytes" (Benzing, 2000). Already, the family has provided numerous subjects to study the mechanisms of drought tolerance and the manner in which some of these same species and others thrive despite their dependence on exceptionally dilute and unconventional sources of mineral nutrients like nitrogen and phosphorus. None of the Florida natives are lithophytes.

About half of the bromeliads—as exemplified by terrestrial *Ananas comosus,* the pineapple—absorb mineral nutrients from the ground with roots. The epiphytes also, of course, need these same substances, but go about obtaining them in truly unusual ways and from pretty unconventional sources. Instead of soil, they rely on decomposing litter (plant debris), rainfall and related washes, captured prey, and plant-feeding ants. The Florida natives, all epiphytes, use all of these sources, except perhaps the ants.

VEGETATIVE STRUCTURE

Bromeliads are soft-bodied by general plant standards, which means that they lack the woody tissue that distinguishes trees and shrubs from herbs. Despite this definition, a fair number of them produce hard, stiff foliage, particularly those types adapted for arid, sunny habitats. These are the species that best fit the name "wild pine."

Evolution has taken the membership of subfamily Tillandsioideae in a most unusual direction. During this progression, the lower half of the body, specifically its root system, has become much diminished, and the upper half, the part that constitutes the shoot system, greatly elaborated. As roots have lost absorptive capacity and proportional mass, the shoot has made up for the loss. Affected most dramatically by this transformation are the species that have come to rely exclusively or nearly so on their densely crowded, scale-shaped leaf trichomes to absorb moisture and nutrients (**CPs 3, 4**).

The typical bromeliad plant is constructed of discrete, serially produced subunits called ramets. Many orchids and other herbaceous perennials that also grow and reproduce in annual increments possess this same kind of architecture (**Figure A**). Each year, at least one ramet is added to what over time will become a physiologically integrated system of repeating vegetative units. Well-established specimens consist of preflowering, reproducing, and spent ramets, the oldest of which are either close to death or have already died.

Each bromeliad ramet amounts to a single determinant shoot, which, except for Spanish moss and a few additional species, is equipped with at least a few roots. Being programmed for limited growth, its growing point or apical meristem produces a certain number of leaves and then an inflorescence (the flower and fruit-bearing portion of the shoot) (**Figure A**).

Immature ramets, or what hobbyists call "pups," arise from lateral buds that are usually located somewhere along the lower half of the parent ramet. With the exception of a few species, none of them Florida natives, they remain dormant until the parent ramet flowers or shortly thereafter. Post-reproductive ramets gradually die over the subsequent two to three years, all the while supplying unused nutrients to younger parts of the plant. So organized, the life of a typical bromeliad is potentially

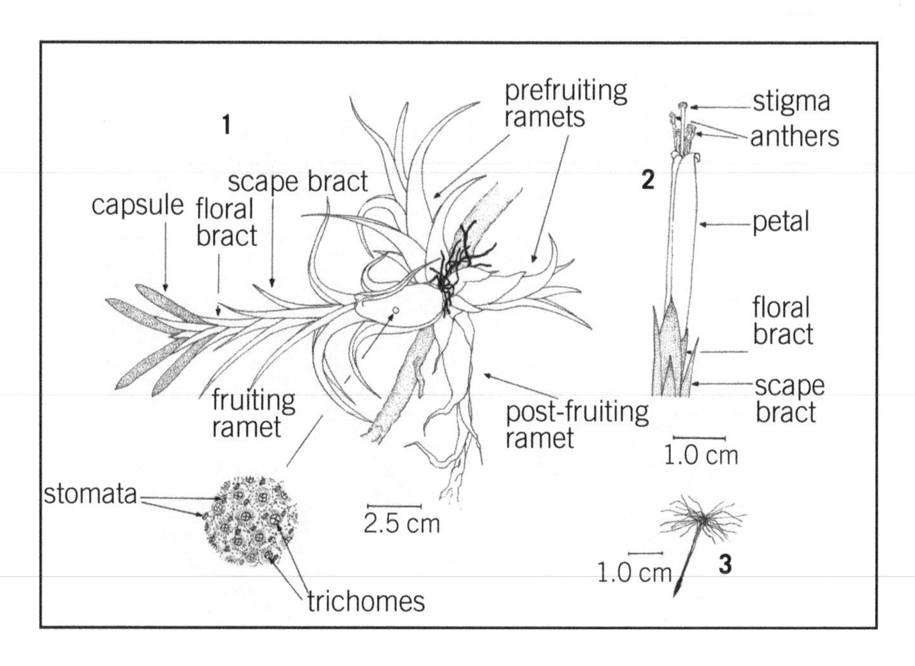

Figure A *Tillandsia paucifolia* specimen showing post-fruiting, fruiting, and prefruiting ramets and details of leaf surface (1), a flower of *Tillandsia setacea* (2), and one of its seeds (3). (D. Benzing drawing)

infinite in the sense that ramets beget ramets in what amounts to an annual rejuvenation process. In reality, of course, accidents of nature assure that life is always finite.

Although all but a few bromeliads exhibit the same basic body plan, ramets vary manyfold in size depending on the species. They can range from only centimeters in any dimension to several meters across and even taller when flowering. The solitary shoot produced by South American *Brocchinia micrantha*—the largest member of the family—becomes palm tree–like. Those of Bolivian *Tillandsia bryoides,* as indicated by its name ("bryo," meaning moss), closely match the largest mosses for size and appearance.

Except for the most anatomically abbreviated of the bromeliads like Spanish moss (*Tillandsia usneoides*), the individual ramet bears a few to several dozen leaves inserted in a tight spiral along a relatively short stem (**CPs 2, 5**). Such shoots are described as rosulate. Ramets with longer stems are considered caulescent, a condition not seen among the Florida natives. Typically short, fibrous, wiry roots originate deep within the stem and emerge between the leaves, most often along the lower parts of ramets.

Once again, the most notable of the species that depart from the family norm are the most diminutive ones because they produce the fewest roots, and sometimes none at all. These are also the bromeliads that represent the endpoint of that evolutionary progression toward shoots capable of performing the absorptive functions usually conducted by roots.

Spanish moss deviates from the pattern exhibited by the bromeliads whose well-established representatives consist of clusters of attached ramets, all dead or moribund except for those representing the youngest two to three generations.

Instead, this species produces a long series of pendant, attached ramets, many of which are alive at the same time. However, each one consists of only three leaves, and if reproductive, terminates growth by producing a solitary, tiny flower (**CP 6**).

Like *Brochinia micrantha* in South America, Florida *Tillandsia utriculata* fruits but once from unusually large, tank-equipped shoots that subsequently die without producing daughter ramets (**CP 5**). Such plants qualify as monocarps (single-fruiting) rather than polycarps (multifruiting), and they require at least a decade's preparation before launching their single life-ending bout of reproduction. Fruit-bearing individuals vary severalfold in size, indicating that something other than a plant's age or mass induces this species to flower.

PHOTOSYNTHESIS AND LEAF FORM

Bromeliads manufacture sugar from carbon dioxide using two distinct biosynthetic pathways, one of which reduces a plant's requirement for moisture more effectively than the other (**Figure B**). CAM-type photosynthesis is the more drought-friendly of the two mechanisms, the more widely occurring C_3 process being the more primitive alternative. Both arrangements occur among the Florida natives, the first type serving most of the species. It's easy to distinguish a CAM from a C_3-equipped bromeliad by its appearance, particularly by the nature of its foliage. Most telling are leaf thickness and texture. Which pathway prevails also influences the conditions under which a species can grow, hence where you will encounter it in nature.

CAM–type photosynthesis promotes drought tolerance by allowing its users to absorb carbon dioxide at night instead of

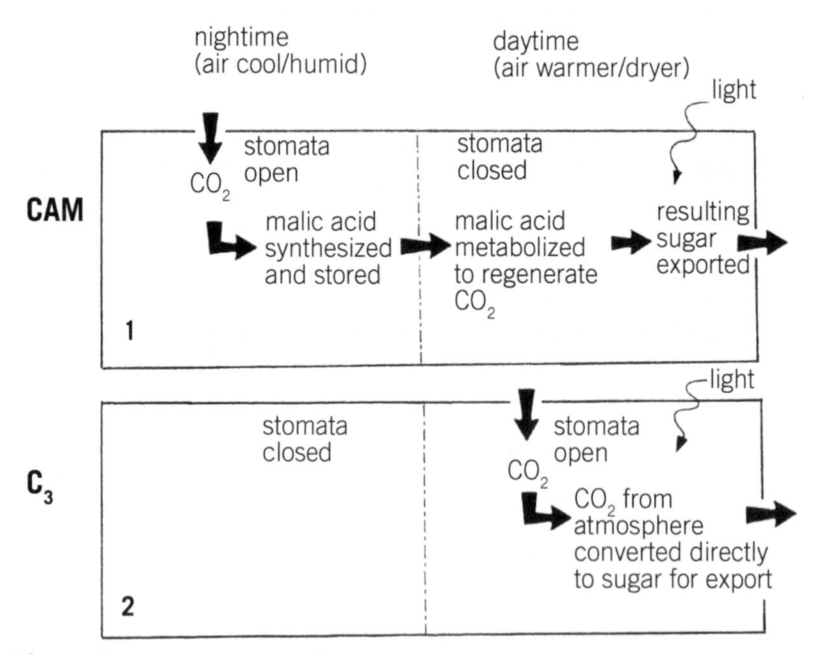

Figure B Schematic depictions of the two photosynthesis pathways that CAM and C_3-type bromeliads use to make sugar. Box 1 illustrates what goes on inside the leaves of CAM-type bromeliads during the night (left side) and day (right side). In essence, CO_2 taken up from the atmosphere at night is fixed twice, first into malic acid and a second time into sugar the following day. C_3-type species utilize only the second, light-dependent step. (D. Benzing drawing)

during the day. Being able to accomplish this essential task after sundown permits a plant to conserve water because nighttime air, having cooled down, has lost much of the evaporative power that it possessed during the day. Consequently, CAM plants don't have to sacrifice as much moisture as their C_3-type counterparts do when they open their stomata (**CP 4, Figure B**). Gas exchange through these tiny pores is bidirectional: while carbon dioxide is entering from the surrounding atmosphere, water vapor exits by a process called transpiration. It's a tradeoff that encumbers all plants that live on land in the sense that doing so requires the expenditure of an often scarce resource, namely moisture, to make food.

C_3-type plants transpire on average about 500 grams of water for every gram of dry matter that they amass through photosynthesis; individuals that employ CAM achieve transpiration ratios as low as 15–20:1. However, plants that employ CAM must pay a high premium for their economical use of water. As experienced horticulturists well know, the C_3-dependent bromeliads grow and mature faster than their more drought-tolerant, but slower-growing, CAM-equipped relatives.

Most of Florida's bromeliads exhibit some leaf succulence, and quite a few of them possess relatively small, compact shoots densely covered with silvery gray trichomes, all three of which signal drought tolerance. On the other hand, species characterized by softer, greener foliage are considerably larger, and they impound substantial amounts of water and litter in proportion to the sizes of their shoots. *Catopsis floribunda,* for example, maintains relatively capacious tanks by virtue of its broad, thin leaves with much-expanded bases (**CP 7**). Conversely, its foliar trichomes are so minute and widely dispersed as to be nearly invisible to the unaided eye. Members of the first description practice CAM, while those of the second fix carbon dioxide via the C_3-type mechanism.

Unable to conduct CAM-type photosynthesis, *C. floribunda* and its kind cannot take up carbon dioxide at night, but given the more or less continuous supply of moisture provided by their leafy impoundments, they don't have to. Being confined to evergreen swamp forest habitats where the air is relatively humid and temperatures are moderated also helps. In contrast, Spanish moss and the other gray, "atmospheric" tillandsias that lack impoundments and face harsher growing conditions maintain better insulated, albeit more modest stores of moisture inside their densely trichome-covered leaves. Members of a third group of Florida bromeliads, for example *Tillandsia bartramii* and *T. simulata,* fall between these extremes, being equipped with small tanks located among the bases of fairly thick, gray leaves (**CPs 8, 9**).

Operation as a CAM-type plant requires more than simply opening the stomata at night and keeping them closed during the day—exactly the opposite of what a plant like *Catopsis floribunda* does. Also unlike what the C_3-type plants do, the CAM types must be able to store the carbon dioxide that they absorb after dark until light is again available to complete its conversion to sugar. Accomplishing this water-conserving feat requires appropriate anatomy and some special biochemistry because carbon dioxide has to be fixed twice rather than just once (**Figure B**).

Sugar cannot be manufactured without using energy derived from sunlight at some point during the food-making process. The CAM-equipped species postpone this step, initially converting the carbon dioxide that they acquire from the atmosphere at night into the common fruit constituent malic acid. Malic acid requires less energy to make than sugar. It also lends itself to storage, but only in a special type of green tissue. This is the same kind of tissue that thickens the stems of cacti and the foliage of dry-growing plants such as the sedums and agaves, in addition to the CAM-type bromeliads.

Around about sunrise, the CAM plant begins to liberate carbon dioxide from the malic acid that it synthesized and stored in those special large green cells the previous night. This deacidification process continues through the day until supplies are exhausted. Unable to escape back into the atmosphere, the carbon dioxide regenerated in this fashion, now with light available, is fed into the same photosynthetic pathway that C_3 plants use more directly to make sugar. Some of the resulting high-energy product must be expended to capture more carbon dioxide from the atmosphere the next night, but enough is left over to fuel plant growth and maintenance.

THE LEAF SCALE

Where and how a bromeliad grows is influenced by more than the shape of its shoot and whether it conducts photosynthesis via the CAM or C_3 mechanism. Equally influential are the services provided by the trichomes or scales that cover its foliage. Small as these appendages may be, they deliver important services. Which of them accrue depends on certain characteristics of the trichomes that deliver them, and these trichome characteristics, in turn, vary with the type of bromeliad.

Epiphytes by nature must compensate for their lack of contact with the ground. Many of the bromeliads counter this problem by assembling what amount to substitute soils in vase-shaped shoots (**CPs 2, 10**). Utilizing these moist accumulations of decomposing litter requires that the adjacent leaves be further modified to operate as roots. This is what the foliar trichomes do, and for the tank bromeliads, they may not do much more.

Bromeliads that lack tanks also substitute trichome-equipped leaves for soil-embedded roots, but they face a greater challenge in doing so. While the tank-equipped individual can draw minerals and moisture from its fabricated organic soil more or less continuously, the atmospheric bromeliad has to act considerably faster and then only sporadically. Everything that an atmospheric bromeliad requires must be captured during storms and shortly afterward before rain-moistened leaf surfaces dry. Trichome operation in this second case often involves a kind of movement that insulates the plant from undue water loss and also damage caused by intense sunlight (**CP 4, Figure C**).

Some of the bromeliads adapted to grow in deep shade, such as *Guzmania monostachia,* feature leaves with trichomes as densely crowded as those borne by the atmospheric types. But when this condition prevails, the shields of their trichomes, whether wet or dry, are rigidly flattened and transparent, rather than capable of flexing upward to reflect excess light like those of the sun-loving species. Trichomes borne by the inhabitants of the darkest microsites, such as *Catopsis floribunda,* scatter so widely and bear such narrow shields that they don't influence a plant's light relations much, if at all (**CP 7, Figure C**).

Figure C illustrates how the atmospheric bromeliads' multifunctional trichome operates. As rain begins to moisten the

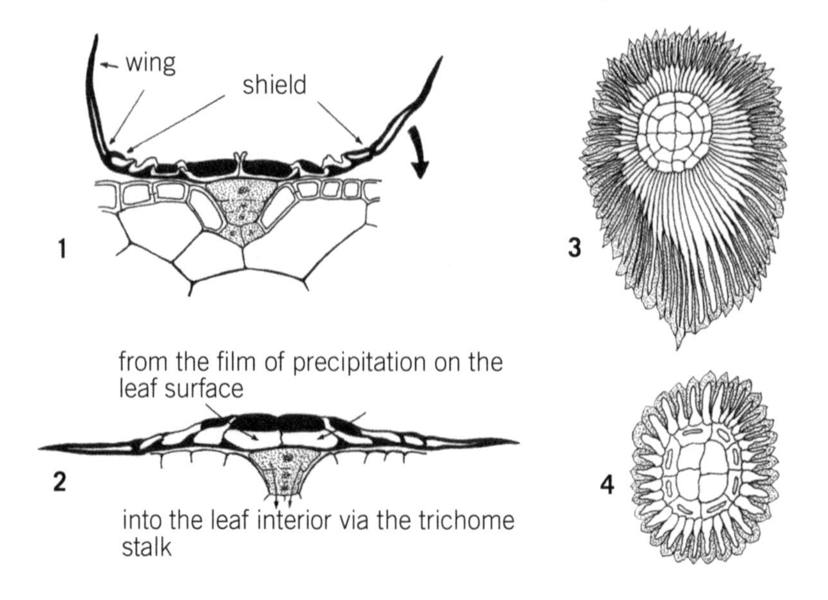

Figure C Foliar trichomes: When dry (1), the wing of the shield of the type of trichome borne by the atmospheric bromeliads flexes upward. When wet (2), it lies flat against the leaf surface and moisture passes into the leaf interior via the trichome stalk. 3 is the top view of the shield of the trichomes featured in 1 and 2. The shield illustrated in 4 is the narrower type produced by the soft-leafed tank formers like *Catopsis floribunda.*

surface of a dry leaf, which for the sake of illustration we'll say bears just one of these appendages, the four, unevenly thick-walled cells that comprise the center of its shield swell upward as they fill with water. Moistening also causes the previously upright wing of the shield to flex downward against the leaf surface. Then, some of the water engorging the four central cells moves down into the trichome's living stalk. From here, it courses into the leaf interior until it eliminates what had been a leaf water deficit.

Nutrients delivered in the precipitation that wets our hypothetical leaf surface reach the interior of the leaf by way of a selective absorption process that begins at the membrane of the uppermost cell of the trichome stalk. Movement through the dead shield cells to reach this location, as for water, is strictly a physical rather than a biological process.

As the storm moves off and our leaf surface begins to dry, the four central cells of the trichome shield pinch shut, returning them to their premoistened, flattened configuration. Thickened upper walls collapse into lower ones precisely sculpted to receive them, thus positioning what amounts to a watertight plug directly over the uppermost stalk cell. Were this final action omitted, moisture could escape from the leaf interior much as a wick exposed to air can empty a bottle of water.

In essence, the trichome, as it serves the dry-growing tillandsias and similarly adapted bromeliads, operates as a one-way valve alternately permitting the plant to relieve water deficits and accumulate nutrients when wet and preventing undue water loss while dry. Because the wing of the shield flexes into its original upright position upon dehydration, its capacity to scatter potentially damaging solar radiation is restored in time to prevent sun scalding.

Tillandsia fasciculata in bloom, Everglades National Park (B. Holst photo)

REPRODUCTION

Most of Florida's bromeliads use birds for pollination. Bright red floral bracts and scentless flowers equipped with long tubular corollas from which the sexual organs protrude discourage all but the largest nectar-seeking insects (**Figure A, CP 11**). *Catopsis nutans* and *Tillandsia usneoides* operate differently, their yellow and chartreuse flowers, respectively, opening late in the day and remaining receptive through the night (**CPs 6, 12**). Powerful sweet fragrances further accord with moths as their primary pollinators. The other two *Catopsis* species probably attract less specialized, day-active insects (**CP 13**). *Tillandsia utriculata* also appears to be adapted for insect visits by, in this case, both diurnal and nocturnal fliers (**CP 14**).

Sexual compatibility varies, as does the need for animals to visit flowers to produce fruits. *Tillandsia recurvata* spontaneously sets seeds with its own pollen, and some additional species may do the same thing less consistently (e.g., *Tillandsia utriculata*). A much-reduced corolla and no rewards for pollinators indicate that ball moss hasn't required assistance to set its fruits for quite some time (**CP 15**).

Members of subfamily Tillandsioideae—hence, all of the bromeliads in Florida—produce small, narrow seeds in three-chambered fruits called capsules (**Figure A, CP 16**). Those of *Catopsis* bear tufts of kinky coma hairs on both ends. A parachute-like device also comprised of hairs on just one end increases the buoyancy of the seeds of *Guzmania* and *Tillandsia.* The tendencies of many of the Florida natives to exhibit distinctly clumped distributions suggest less mobility than might be expected for propagules that appear so elaborately equipped for wind dispersal.

FLORIDA'S BROMELIADS VERSUS ITS OTHER EPIPHYTES

Epiphytism has emerged repeatedly across the plant kingdom. Groups beyond Bromeliaceae that also engage heavily in this peculiar lifestyle include the ferns, as well as the flowering plant families to which the aroids, cacti, orchids, and the African violets belong, more than 80 in all. Florida's epiphytes hail from about a dozen and a half of these groups, with Orchidaceae contributing the most species. Bromeliaceae ranks third, and the ferns come in second, increasing the total by about twenty-five more. (Benzing, 1990). Altogether, the state supports more than 80 kinds of epiphytes, including a couple of which that grow on the ground so often they barely fit the definition.

It's not surprising that numerically the orchids and ferns rank first and second among Florida's epiphytes. Much the same pattern prevails worldwide. About one third of the roughly 9000 ferns grow on woody hosts, and the orchids account for more than half of the roughly 25,000 free-living epiphytes. So with only about 1500 of its members engaged in this lifestyle, Bromeliaceae shouldn't be expected to dominate Florida's epiphyte flora. Biological importance is another matter, however. Virtually everywhere across the American tropics, the bromeliads exceed their epiphytic fern and orchid companions many times over for sheer biomass and also often for numbers of individuals. This relationship would apply in Florida even if Spanish moss were the only member of its family included in the comparison.

Most epiphytes colonize multiple kinds of hosts, but rarely every possibility, and often for obvious reasons. Florida's bromeliads utilize more kinds of supports than any of the state's other indigenous epiphytes. Those hosts used most frequently—Carolina ash, cypress, and live oak—for example, possess rough, stable barks borne on similarly long-lived trunks and branches. Quite a few of these hosts inhabit wetlands, where humidity further favors epiphyte success. Trees with smooth or unstable barks, such as gumbo limbo and the strangling figs, seldom accommodate epiphytes beyond the occasional bromeliad on gumbo limbo. Claims that secreted plant toxins further determine host preferences remain largely speculative.

Adaptations for epiphytism exhibited among the Florida epiphytes vary by family, especially when it comes to roots. The state's mistletoes lack these organs altogether, relying instead on a host-penetrating device called a haustorium. All of the free-living types except Spanish moss continue to employ roots, although for the bromeliads that retain them little is gained other than mechanical anchorage. It's the orchids that best illustrate how much an epiphyte's root system can do beyond attaching it to its host.

The roots of orchids haven't undergone the diminution that those of the bromeliads experienced as their terrestrial ancestors abandoned soil in favor of bark (Benzing,1990). Orchid roots not only remain undiminished in mass in proportion to the shoots that produce them, they exhibit considerable anatomical refinement. Perhaps most important for operation above ground is the presence of a multilayered spongy epidermis called a velamen, and just below it, a layer of photosynthetic tissue, which in turn surrounds the vascularized root core. So equipped, the roots of the epiphytic orchids operate much like the trichome-bearing leaves of the bromeliads by absorbing both water and mineral nutrients, but they also supplement foliage as makers of food. The much thinner, wiry bromeliad root lacks these special features, as do those of the ferns.

Unable to invade and tap a host's vascular system as the

mistletoes do, the free-living epiphytes must exploit other options. Given the obscurity of most of their sources of mineral nutrients, it's no wonder that the true epiphytes so often inspire the designation "air plant." Particularly enigmatic in this score are the bromeliads like ball moss that thrive on telephone wires and other similarly sterile anchorages (**CP 17**).

Quite a few bromeliads obtain nutrients from ants by providing nest sites inside onion-shaped shoots or, if tank formers, between leaves that have become too old to impound moisture. *Tillandsia paucifolia* approaches the first condition, although in Florida, ants use it only sporadically (**CP 3, Figure A**). A group of more specialized bromeliads native to other regions, along with dozens of epiphytes representing about ten more families, regularly harbor ant nests in more elaborate, leaf- or stem-derived accommodations called myrmecodomatia.

Whether *T. paucifolia* or similarly bulbous *T. balbisiana* (**CP 18**) receive significant nutritional supplements from their casual insect users in Florida isn't known. Local ants more faithfully patrol the developing inflorescences of *Tillandsia balbisiana* as described below, but they mostly nest elsewhere.

Florida's bromeliads demonstrate the fundamentally tropical nature of the epiphytes by the fact that of the sixteen species, only three also grow in adjacent, colder states. Spanish moss thrives from coastal Virginia all of the way south to central Argentina, sometimes growing on rocks as a lithophyte. Why it survives temperatures that kill its close relatives isn't known.

Except for endemic *Tillandsia simulata,* all of Florida's bromeliads range far southward, most of them at least into Central America. *Catopsis berteroniana* holds the record, penetrating into South America to east central Brazil, and several of the tillandsias reach Bolivia or Peru. The extra-Florida distributions of the state's epiphytic ferns and orchids reveal similar tropical affinities.

Mention has already been made of the importance of absorbing trichomes and leafy tanks to the Florida bromeliads. Other adaptations also identified as favoring epiphytism are leaves that store water in special tissues. Numerous non-bromeliad epiphytes, including many of the Florida orchids, also employ CAM-type photosynthesis, and relatively desiccation-resistant foliage is almost the rule. The state's single drought-deciduous (leaves shed during the dry season) orchid (*Crytopodium punctatum*) employs its massive trash-basket root system to intercept litter, as do some of the ferns.

Additional requirements for epiphytism concern reproduction, which as noted above for the bromeliads, include seeds air-buoyant enough to disperse from tree to tree. Birds and digestion-resistant seeds packaged in fleshy fruits accomplish the same thing for members of subfamily Bromelioideae and a host of epiphytes in other families like Araceae and Cactaceae. The orchids produce even smaller wind-dispersed seeds than the bromeliads of subfamily Tillandsioideae, and fern spores are smaller still. Florida's single native epiphytic cactus, as well as the strangling figs, ripen soft-walled fruits containing hard-coated seeds capable of passing undamaged through the guts of birds.

DO BROMELIADS EVER HARM THEIR HOSTS?

An epiphyte doesn't have to be a parasite—in effect, a mistletoe—to injure its host. Florida's bromeliads illustrate just about all of the possible ways that nonparasitic plants can harm trees—some of which mimic, or at least suggest, parasitism, while others do not.

Branches felled by the added weight of large festoons of Spanish moss require no speculation as to cause. More often, however, the adversely affected host remains intact, but is probably too heavily shaded by dense trusses of this same bromeliad to grow normally. Several additional species, most notably ball moss and the largest of the tank formers, *Tillandsia utriculata,* similarly challenge trees by overburdening or heavily shading them.

Free-living epiphytes can also injure their hosts by compromising the nutrient cycles upon which all large-bodied, long-lived plants depend. It's a subtle process, but it's not parasitism. The appropriate label for this phenomenon is "nutritional piracy."

To be a nutritional pirate, a bromeliad, or any other kind of free-living epiphyte, need only co-opt nutrients as they normally cycle between a host and the soil from which that tree or shrub obtains them (**Figure D**). How much nutritional deprivation a tree infested with epiphytes experiences depends on a variety of site-specific circumstances. It's not the existence of nutritional piracy that's in question here, only the seriousness of its consequences for hosts. Whatever the impact, the process works in the following way.

Trees and shrubs lose nutrient capital whenever they disperse seeds or jettison old foliage and other spent organs, all of which contain significant amounts of nitrogen, phosphorus, and the other nutrient mineral elements. Deficits caused by these activities tend to be temporary, however, the losses being recoverable.

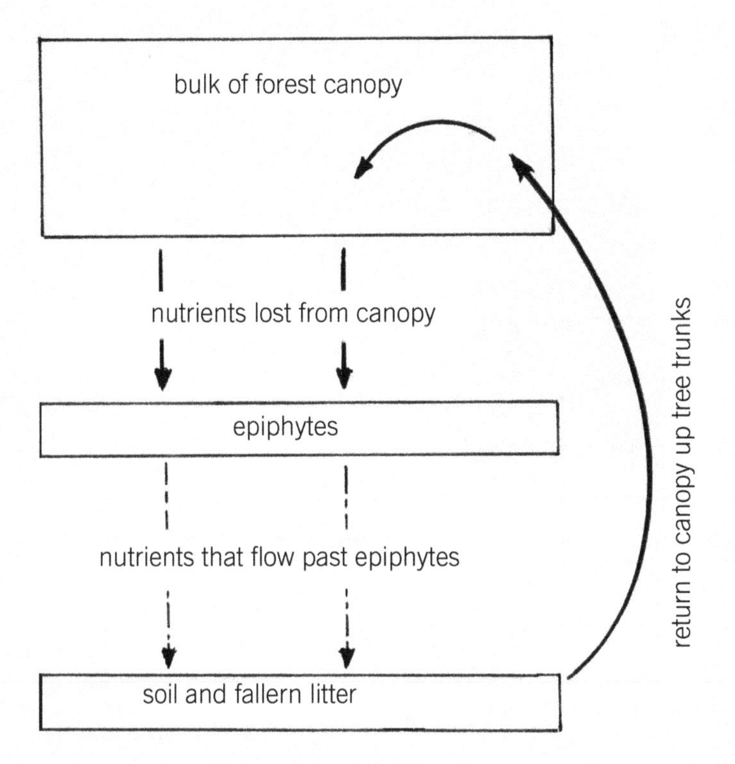

Figure D A schematic representation of a phenomenon known as nutritional piracy that describes how free-living epiphytes deny their hosts opportunity to recover key mineral nutrients by extracting them from litter and canopy washes as part of the biogeochemical cycling process in woodland ecosystems. See text for additional details. (D. Benzing drawing)

Litter eventually decays, releasing its mineral constituents. If this process occurs on the ground within reach of the roots of the trees that shed the litter, what was lost can be retrieved. It's all part of an important ecosystem-wide process described as biogeochemical cycling.

Biogeochemical cycles tend to be tight rather than leaky in mature, stable forests compared to regenerating or disturbed ones. But efficient recapture is most crucial for plant welfare where soils are thin or infertile. Trees growing at such locations risk experiencing serious nutritional stress should they have to replace, from such deficient sources, substantial amounts of elements like phosphorus or potassium lost to epiphytes, or for that matter, to any agency.

Bromeliads interdict by two routes the streams of nutrients that move between their woody supports and the soils in which those trees root (**Figure D**). Either way, a host experiences the same impact; it can't recover lost nutrients as effectively as would be possible were its crown epiphyte-free. The relatively slow-growth characteristic of most epiphytes assures that whatever resources get co-opted from a host by a dense colony of bromeliads remains unavailable for recycling for years to come.

Tank bromeliads subsist primarily on nutrients derived from litter as it decomposes within their leafy impoundments. Lacking this option, Spanish moss and its kind survive largely on what they extract from precipitation and the washes that result from precipitation. Much of what is useful for epiphytes in this second source represents substances leached from still intact plant parts higher up in the crowns of their hosts.

Oaks, among the many kinds of trees favored by Florida's bromeliads, seem to display most dramatically the inverse relationships between epiphyte infestation and host health. In extreme cases, dead supports continue to accommodate massive colonies of Spanish moss, although whether or not nutritional piracy contributed to the demise of these individuals cannot be confirmed. Quite possibly, the additional light that can pass through a leafless crown simply ramped up bromeliad photosynthesis. Or perhaps pathogens were involved, conceivably rendered more virulent for a host already weakened by a heavy load of epiphytes. Whatever the cause or causes, trees that bear

Heavy growth of *Tillandsia setacea* on an oak, Deer Prairie Creek Preserve
(B. Holst photo)

the densest infestations of bromeliads often appear worse off for that association.

FLORIDA AS A HABITAT FOR BROMELIADS

Florida hosts some of the most botanically rich ecosystems

anywhere within the contiguous United States. Home to fully 3500 species of higher plants, it easily qualifies as a horticulturist's paradise. Several factors account for this distinction, most importantly the state's warm temperate to subtropical latitudes, and, as a relatively narrow peninsula, its proximity to climate-moderating seas.

So situated, Florida provides favorable growing conditions for relatively frost-hardy biota in the north and for truly tropical species in the south. In effect, it constitutes a botanical crossroads, populated by colonists delivered by winds and ocean currents from the Caribbean at one end and by air currents and other agents from the rest of North America at the other.

Having warm water on three sides also assures Florida more precipitation than typically falls at comparable latitudes elsewhere, for example, in decidedly drier northern Mexico. Historic marine influence is evident in the sandy soils over coral rock that characterizes much of peninsular Florida. Both of these layers were deposited during the state's repeated submersions as global sea levels fluctuated more than 250 feet during just the past several hundred thousand years. Shells produced by the same kinds of organisms that still flourish offshore today occur inland except on the highest parts of Florida's central ridge system. Only the uppermost reaches of this rocky spine remained unaffected by melting and accreting continental glaciers far to the north and south.

During the most recent 1.5 million years, Florida's terrestrial biota has occupied both considerably more and far less land area than is colonized now; only about 10–15% of the peninsula's current extent has escaped intruding seawater for more than the past 110,000 years. The southernmost quarter of the state was inundated as little as 3000–5000 years ago. Consequently, the home for Florida's certifiably tropical flora, that part that includes most of the bromeliads, is its youngest portion.

It's likely that several of the southernmost bromeliads of Florida, like *Catopsis nutans* and *Tillandsia pruinosa,* repeatedly died out and recolonized the state as global climate fluctuated. Nearby Bahamian natives like *Tillandsia bulbosa* could well have been Florida residents during one or more of the interglacial phases that preceded this one. Perhaps it lived here even earlier during the one that that we enjoy today.

Geology also contributes to the state's exceptional floristic diversity, including what it does to support epiphytes. Limestone erosion has produced countless sinkholes, many connected by subterranean channels, the result being a patchwork of wetlands of varying water chemistry, forest type, and hospitality for bromeliads. More conspicuous evidence of the powerful dissolving capacity of rainwater, the aptly labeled pinnacle rock lies exposed through much of Collier and adjacent counties in extreme south Florida. Thick accumulations of humus overlaying this substrate at some sites support substantial populations of a number of otherwise epiphytic species like the whisk fern *Psilotum nudum.*

Quite recently by geologic standards, humans initiated their catastrophic influence on Florida's natural landscapes. Native Americans arrived at least 10,000 years ago, but until corn, beans, and some other crops domesticated in Mesoamerica were adopted, the native vegetation probably didn't change much. Europeans made a bigger impact by introducing more destructive farming and logging practices. Degradation reached even greater heights as development intensified, particularly activities like phosphate rock mining, citrus culture, and dredging designed to lower water tables.

Much of what remained as relatively unaltered pine and oak savannas and scrub cypress forests fell victim, at the end of World War II, to devastating wildfires fostered by drier conditions promoted by massive engineered drainage. Today, almost no mature cypress woodland survives, and what primary mangrove forest persists occurs largely within state and federal park boundaries. All of these early assaults—plus hurricanes, saltwater intrusion, fertilizer runoff, and invasions by alien plant species—have left much of Florida's rural landscape quite altered from its presettlement condition.

Despite the destructive interventions by humans, bromeliads, particularly *Tillandsia usneoides* and *T. recurvata,* continue to inhabit large swaths of scrub-dominated uplands, pine flatwoods, palmetto/pine savannas, and cypress and hardwood forests. Many urban landscapes have proved almost as accommodating to the same two species. Much less remains of the populations of the other bromeliads that once inhabited the mangrove and coastal strand communities that extended down the west coast from Tampa Bay through the Florida Keys and up into Miami-Dade County.

Florida's surviving bromeliads range from abundant and widespread to narrowly distributed and rare, the latter being those species restricted by frost-sensitivity to the extreme south (e.g., *Catopsis* spp. and *Guzmania monostachia*).

Tillandsia fasciculata and several of its relatives once grew abundantly on low-growing hosts, and sometimes even on the sandy ground below in what were extensive coastal strand communities before development came along. Species that exhibit this kind of ecological duality with sufficient regularity, or begin life about as successfully on either bark or soil, qualify as facultative instead of obligate epiphytes. Florida bromeliads, for the most part, are obligate, dying within weeks if stranded on any but the most thoroughly drained soils. The same scale-shaped trichomes that work so well under drier conditions, if kept wet for too long, suffocate the over-watered plant by blocking gas exchange through its stomata (**CP 4, Figure C**).

Drought- and shade-tolerance continue to influence the distribution of the Florida bromeliads. Species with scruffy, silvery leaf surfaces still grow mostly exposed to near or full sun, whereas those that possess thinner, smoother foliage and vase-shaped shoots occupy more shaded, humid locations. Several of the larger tank-forming bromeliads also continue to demonstrate the flexibility of the individual plant. Specimens of *Tillandsia utriculata* grown in unfiltered sunlight, for example, display compact, gray-reflective, stiff-leafed shoots, whereas their shaded counterparts exhibit a greener, more lax-leafed form (**CPs 19, 20**).

THE IMPORTANCE OF BROMELIADS IN FLORIDA HABITATS

Plants contribute a variety of products and services to other nearby organisms. The nature and importance of these contributions depend on all sorts of things, including who is offering what, when, and where. Discrete items like food and shelter quickly come to mind; some of the services aren't so easily identified.

Epiphytes also influence a number of ecosystem-wide processes like nutrient cycling and whole-system photosynthesis. Whether a specific species of epiphyte, or any other type of resident, is a major or minor player in one or more of these roles depends on its abundance, physical location, and basic botanical nature. The tank bromeliads can be inordinately important to their immediate neighbors as well as to the broader biological community to which they belong.

Mention has already been made of the rewards that the Florida bromeliads produce for flower-visiting birds and insects. Recall also how these bromeliads and the other kinds of epiphytes, if present in large enough numbers and under the right circumstances, can seriously deprive their hosts of key nutrients.

The rest of this discussion focuses on the tank bromeliads as providers of high-quality habitat in the form of what are called "phytotelms" or "phytotelmata" (Benzing, 2000; **CPs 5, 21**). A phytotelm (singular—also spelled "phytotelma") is a plant-based, water-filled cavity that can be an important resource for animals of many descriptions. Other kinds of phytotelms include water-filled holes in tree trunks and the inflated leaf bases of certain members of the banana family.

The vertebrate and invertebrate animals and microbes that populate bromeliad phytotelms qualify as mutualists to the extent that they promote plant welfare, especially nutritional well-being. In return, the plant provides hiding, feeding, and breeding opportunities. Only *Catopsis berteroniana* among the Florida species acts more one-sidedly by killing rather than nurturing some of the animals that end up in its leafy impoundments (Frank and O'Meara, 1984). How this plant operates as a botanical carnivore is described below.

Numerous surveys have documented myriad worms, insects and other arthropods, and amphibians living in the water-filled leaf bases of bromeliads examined at sites from Brazil to Florida (Laessle, 1961; Benzing, 2000). Not all these inhabitants feed on detritus, however. Rather, some of them eat each other;

and the members of a third group are just visitors attempting to avoid this fate, otherwise conducting their lives elsewhere. The microbes present are even more opportunistic, thriving just about anywhere enough moist debris collects to constitute food. Whatever the relationship, all of this biological activity so close to absorptive foliage relaxes the tank bromeliad's need for roots except as anchorage.

Some of the animals that inhabit the watery impoundments provided by bromeliads display characteristics that suggest absolute dependence on their botanical benefactors. Diverse species exhibit behaviors and structures suggestive of long histories of associations with leafy tanks. Among these users are several diving beetles belonging to the genera *Aglymbus* and *Copelatus,* a number of ostracods, the larva of at least one Brazilian damselfly, and some Jamaican crabs with close, marine-dwelling relatives.

A substantial number of small Mesoamerican frogs breed exclusively in bromeliad phytotelms, although not in Florida. Enabling characteristics include clutches comprised of just a few eggs provisioned with oversized yokes to support rapid development in small pools that ordinarily provide little food for tadpoles. A collection of equally dependent, bromeliad-dwelling salamanders native to Mexico and Central America feature slender limbs and trunks presumably evolved to improve mobility within the narrow spaces created by closely overlapped leaf bases.

Frogs are frequent, although not obligatory, users of bromeliads in Florida. Cuban tree frogs (*Osteopilus septentrionalis*) can be regular occupants by day, often returning to the same plant after each night's foraging. *Hyla cinera,* the green tree frog, also employs bromeliads for daytime shelter.

Florida's atmospheric bromeliads probably benefit co-occurring animals less than their tank-forming relatives do. Spanish moss could be an exception, considering its abundance, wide occurrence, and appeal as a nest-building material for certain birds.

Plant eaters may pay less attention to the bromeliads than to many nearby plants. Relatively low concentrations of nitrogen and other key nutrients in bromeliad foliage, especially that produced by the atmospheric species, may explain why these organs often seem less damaged by herbivores. Fruits, seeds, and emerging inflorescences attract more attention, probably because they score higher for key nutrients. An exception is a recently introduced weevil that attacks the vegetative stem of the larger species as described below. In general, however, low quality as forage may be one of the advantages of being an epiphytic bromeliad.

How much the bromeliads of Florida enhance the carrying capacities and biodiversities of the state's woodlands warrants closer scrutiny. The larger-bodied tank formers certainly extend in time the availability of moisture for canopy-based fauna compared to the epiphytic ferns and mosses that often accompany them. Those highly eroded, calcareous soils that support many of south Florida's wetland forests can become dry enough in late winter and spring that any tank bromeliads present will become vital resources for thirsty animals.

Guzmania monostachia sometimes occurs at densities comparable to those recorded in several locations in South America, where bromeliads with similar shoot architecture represented "hot spots"—sites of especially high invertebrate abundance and variety (**CP 10**). Finally, importance may fluctuate by season. Quite possibly, those massive colonies of *Tillandsia* *fasciculata* on pond cypress in the Big Cypress Swamp of south Florida become refuges for terrestrial animals displaced by high water each summer (**CP 2**).

CONSERVATION STATUS AND THREATS

Habitat destruction and plant collectors rank first and second as most responsible for the tenuous futures of many of Florida's bromeliads (Ward, 1979). Narrow distributions further exacerbate the threat for natives like *Catopsis nutans* and *Tillandsia pruinosa.* Chance of extirpation by a few (and ultimately even a single) catastrophic events like storms or fires rises as the area inhabited diminishes.

Certain aspects of the epiphytic way of life heighten vulnerability even more. Epiphytes face special challenges because of how and where they grow. Maintaining adequate nutrition and hydration usually requires more specialized structure and function than for plants that root in the ground. Moreover, bark isn't a particularly hospitable medium for anchorage. Especially problematic is its relatively short life. Twigs typically break off within a couple of years, as do the tissues that make up the outer layers of larger stems and trunks.

Life for the epiphyte is made still more precarious by how the dual threats posed by short-lived rooting media and scarce resources each reinforce the effects of the other. The more time required by a plant to reproduce relative to the duration of its rooting medium, the more likely it will die having produced no offspring. The same can be said for populations and entire species of epiphytes. You might ask then, why hasn't evolution favored faster growth among the epiphytes? It hasn't because it can't.

Epiphytes, especially the most stress-adapted types like the atmospheric bromeliads, mature slowly and for good reason. Doing so diminishes plant demand for vital resources like moisture and key nutrients, effectively reducing the impact of scarcity on survivorship. For example, *Tillandsia pruinosa* and *T. paucifolia* in Florida, despite their modest sizes, require six to ten post-germination years to flower for the first time (**CPs 1, 3**). Growing faster would cause them to outstrip their meager supplies of key resources, decreasing prospects for reproduction and survival accordingly.

Being better supplied with moisture and nutrients owing to their more favorable architecture, many of the tank formers mature in less time. Exceptions occur at both extremes, however. Atmospheric *Tillandsia usneoides* grows surprisingly fast, whereas more than a dozen years may pass before a tank-bearing and monocarpic *T. utriculata* specimen produces its single end-of-life crop of seeds. Speculation about why Spanish moss grows faster than its lifestyle predicts is considered in the following descriptions of the Florida species and their hybrids.

Epiphyte success is further constrained by high mortality, much of it unrelated to shortages of moisture, or of nutrients, or because of short-lived anchorages. Multi-year observations of *Tillandsia paucifolia* growing on dwarfed cypress trees in Collier County demonstrated just how difficult maintaining populations can be for an atmospheric bromeliad (Benzing, 1981). Only 2–4% of the seeds that had been glued to the same trees that supported naturally established individuals yielded seedlings that were still alive one year later.

Five or more additional years were required for the year-old survivors to flower, and few of these individuals lived long enough to do so. Add to the calculation how few of the seeds dispersed by an epiphyte end up lodged on something suitable for establishment, and you can appreciate why plants that engage in this lifestyle might be more vulnerable to extirpation than plants in general. It's also worth noting that most of the fruiting individuals were producing only one or two capsules, each containing only about 150 seeds.

Some of Florida's bromeliads, particularly *Tillandsia recurvata* and *T. usneoides,* range widely enough and thrive on enough kinds of supports to avoid extirpation by any natural event short of a major meteor impact or catastrophic climate change. However, additional species whose high numbers and extensive ranges also suggest pretty substantial immunity may not actually be so secure. If you factor in the potential effects of the recently introduced Mexican weevil, *Metamasius callizona,* the likelihood that large-bodied Florida natives like *Tillandsia fasciculata* and *T. utriculata* have bright futures in Florida isn't so great (Frank, 1999–2000; **CP 22**).

Ongoing investigations indeed confirm that *Tillandsia fasciculata* and *T. utriculata* are the most weevil-threatened of the state's bromeliads. Ten other species exhibit lower rates of attack, while *Tillandsia bartramii, T. recurvata, T. setacea,* and *T. usneoides* never become large enough to attract egg-laying females. The larger the bromeliad shoot, the more weevil larvae it can nourish. Whereas a single large specimen of *T. utriculata* often supports six, or occasionally even more, pupating larvae, fewer individuals (often just one) infest the more diminutive bromeliads.

Vulnerability varies for at least one additional reason. Being monocarpic, hence capable of reproducing just once, and then exclusively by seed, makes large-bodied *Tillandsia utriculata* all the more threatened. *Tillandsia fasciculata,* the second largest of the Florida natives and also a favored food of the weevil *Metamasius callizona,* experiences less mortality because it produces offshoots in addition to seeds. Survival is possible because even individuals infested multiple times can sometimes manage to propagate via one or more undamaged lateral buds. A second member of genus *Metamasius* native to Florida also attacks bromeliads, primarily the smaller ones like *T. setacea,* but too sporadically to threaten entire populations (**CP 23**).

Metamasius callizona is too new to Florida to know what its ultimate effect will be. Almost certainly, the future for *Tillandsia utriculata* would be brighter if it lacked relatives that sustain the same herbivore. This being the case, weevils are more likely to remain numerous enough to search out even the most isolated of the remaining *T. utriculata* plants and destroy them. Consequently, its survival may hinge on the extent to which the state's population includes reproductively viable colonies in habitats unsuitable for the weevil. Should the tachinid fly that parasitizes *M. callizona* in Central America perform similarly following its controlled release in Florida, prospects will brighten substantially.

Concerns about public health could also influence the fates of Florida's tank bromeliads sometime in the future. Mosquitoes that breed in bromeliad tanks might someday transmit malaria or dengue fever, as has occurred at some other locations. Problems with disease-vectoring mosquitoes have been serious enough in several Latin American countries to prompt spraying for larvae or the plants that support them. No comparable threats exist in Florida today, but this could change. The combined effects of global warming and increasing numbers of visitors to the state from regions plagued by insect-borne maladies could make a difference. Concern about nuisance mosquitoes already makes tank bromeliads unwelcome neighbors to many Floridians.

NATURALIZED BROMELIADS

Globalization involves more than just the blending of cultures and linking of national economies. No less a part of the process is an ongoing, human-assisted mixing of our planet's biota. Many animals and plants, indeed even pathogens like HIV, no longer remain confined to their native ranges. Especially adaptable species like the English sparrow and dandelion have effectively "escaped" to take up permanent residences well beyond their ancestral African, Asian, or European homes. Some of the most aggressive of these invaders, such as the agricultural weeds, owe their success to high suitability for widely available habitats. Others have succeeded in part by outdistancing the predators and pathogens that deny them comparable dominance where they came from.

Capacity to establish reproducing populations in alien territory, essentially to become "naturalized," is not a widely shared plant trait. A number of identified characteristics, and probably many others not yet recognized, predispose some species for this kind of performance more than others. The most common type of botanical colonizer is short-lived, herbaceous, and prolific—essentially a weed by nature. These are the species best suited to perform well in ephemeral, or what ecologists consider disturbed, habitats.

Being unusually fecund and able to mature in months rather than years, the annual weeds excel as exploiters of waste areas and roadsides where vehicular or foot traffic or some other agency keeps soils too disturbed to foster success for longer-lived species characterized by slower growth but ultimately taller, more shade-casting stature. Those invaders that succeed despite perennial life cycles and massive woody bodies possess unusual attributes, and they can be even more troublesome than their shorter-lived counterparts. Examples in Florida include the shrubby Brazilian pepper and the Australian punk tree.

Epiphytes rank among the least disposed of the many kinds of plants to colonize alien territory, even regions endowed with growing conditions apparently much like those at home. Florida's exceptions amount to a couple of South American hemi-epiphytic figs whose recent naturalizations through much of the southern half of the peninsula are probably attributable to the production of abundant bird-dispersed seeds and what is essentially a terrestrial character. The epiphytic stages of these plants last only until long roots reach the ground, after which a host is dispensable. At this point, the woody hemi-epiphyte may become a freestanding tree.

Dozens of nonnative bromeliads have already had considerable opportunity to naturalize in Florida, but so far few, and possibly none, have succeeded. Dozens of ornamental, tank-forming species and hybrids representing *Aechmea, Billbergia, Neoregelia,* and related genera of subfamily Bromeliaceae after became popular choices for outdoor culture in southern Florida decades ago.

Abandoned or escaped clones have proven that a number of imported bromeliads can persist in waste areas for decades, but this is as far as things seem to have gone as of today. Dispersal shouldn't be a problem for berry-producing bromeliads capable of fruiting in the wild. Brazilian pepper, that notoriously successful invader in Florida, produces seeds of about the same size and packages them within similarly constructed edible berries.

Past failure doesn't preclude future success, however. Sometimes, plants invade new territory slowly, perhaps first having to fine-tune existing genotypes to better cope with local conditions. Any exotic bromeliad following this trajectory in Florida will require considerable time to do so given the multiple years needed just to achieve sexual maturity. Depressed fruiting in alien habitats may further slow, and perhaps prevent, naturalization for many of the more widely cultivated candidates. Strict self-incompatibility definitely impedes naturalization in some cases (e.g., widely cultivated Brazilian *Billbergia pyramidalis*).

At least one exotic bromeliad is reproducing sexually in an ocean-side park in Broward County, according to Pemberton and Liu (2007). At issue is *Tillandsia ionantha,* a frequently cultivated epiphyte indigenous to southern Mexico and much of Central America. Despite being self-sterile and bird-pollinated, its Florida colony already consists of more than sixty adults, some fruiting, and seedlings on more than thirty Australian pine trees. The ultimate challenge to this normally tropical bromeliad in Florida will come the next time the southern part of the state experiences one of its infrequent, but definitely not rare, deep freezes.

CULTIVATING FLORIDA'S NATIVES

Florida's native bromeliads differ by aesthetic value, conservation status, and ease of culture. *Tillandsia recurvata* and *T. setacea,* and some others of comparably low appeal, seldom find their way into managed landscapes or indoor collections. Moreover, some of the most attractive species languish or die (e.g., *Catopsis berteroniana*) unless provided exacting culture. Being protected by Florida law, and in some cases truly threatened with extirpation, none of the more desirable choices should be collected anyway.

In the final analysis, growers bent on cultivating a Florida native or two should focus on the most common of the state's tillandsias (Plever and Brehm, 2003). Broader ambitions can be satisfied without reducing already diminished wild populations by patronizing reputable nurseries, meaning establishments known to propagate what they offer for sale rather than stocking their inventories with wild-collected plants.

Tillandsia fasciculata with its candelabra-shaped inflorescence densely packed with cardinal red bracts tops the list for commercial exploitation (**CP 24**). Several somewhat less attractive natives like *T. balbisiana* and *T. flexuosa* often accompany it in cultivation (**CPs 18, 25**). Maintaining most of these bromeliads is easy if the conditions provided parallel those that prevail in the wild. Beginning with seeds is a more daunting task, but, again, possible if treatments mimic conditions in nature.

Epiphytes require irrigation and fertilization tailored to match special needs that vary with the species. However, some other procedures apply across the board. Seldom are formulations containing large fractions of humus-rich soil or additives like peat moss appropriate media for epiphytes. Anything that stays moist too long can spell trouble, and it's sure to bring unmitigated disaster for the dry-growing, atmospheric bromeliads for the reason described on page 41 for why dislodged plants usually die on the ground. Good air circulation is another imperative: make sure to provide it, and keep your plants well separated to receive its benefits.

Nature can provide most of what your plants require if you live in the right place. Where climates are accommodating and trees are available, use them as substrates. Elsewhere, a piece of driftwood, a slab of cork or of tree fern roots, or a section cut from a cypress trunk or some similarly rough-barked tree, will suffice (**CP 26**).

Take care to choose chemically inert binders like insulated

wire or nylon fishing line for mounting—avoid copper and zinc, to which bromeliads are exceptionally sensitive. Be sure to position your specimens as they grow in nature. Stand the tank formers upright so they can retain water. The atmospherics do well regardless of orientation. Small adults and seedlings will be fine if secured to their rooting media with Elmer's glue or some similarly durable, nontoxic adhesive.

Keep in mind that those epiphytes that root on small twigs or bark free of mosses and lichens experience desertlike conditions in nature. Most vulnerable to overwatering are these species and the others that grow where nonabsorbent media, high exposure to light, and moving air assure that they experience rapid wet-dry cycles as a matter of course.

The water-impounding bromeliads tolerate greater humidity than the atmospherics, but they may still fail if potted in mixtures that provide the amounts of moisture that many terrestrial plants require. Avoid this problem by using coarse, porous media if specimens must be container-grown. But it is preferable to instead mount them exposed on some kind of quick-drying substrate, as detailed above.

However tank bromeliads are grown, it is important to keep the tanks both debris-free and filled with fresh water. Regular flushing, as precipitation does in nature, prevents injury to young foliage and emerging inflorescences caused by salts concentrated by evaporation. It also reduces chances that any overlooked organic material will putrefy. Use the cleanest irrigation water available to further reduce these possibilities.

Nutrients should be provided frequently and sparingly. One fertilizer mix formulated specifically for tillandsias contains nitrogen, phosphorus, and potassium as a 17:8:22 mixture. Standard commercial products like Rapid Grow and Miracle-

Gro will do if something more specific for bromeliads isn't handy. Micronutrients usually aren't needed. Solutions diluted to no more than 25% of recommended strength should be sprayed over the entire plant once every two to four weeks. More frequent applications delivered at even lower strengths come closer to conditions experienced in nature. Concentrations suitable for most soil-rooted plants can damage many of the epiphytic bromeliads.

Irrigate and apply fertilizer solutions only when light intensity is low. Wetted leaf surfaces compromise the screening effect that dry, reflective trichomes provide to underlying, delicate green tissues (**CP 4**). Undiminished sunlight can fatally scald a moistened shoot, especially if it's been growing in shade. Remember also that the thicknesses of the leaves and the densities and reflectivities of trichome layers reveal whether bromeliads require some shade and relatively high humidity (if thin-leafed and lightly-trichomed) or more intense sunlight and drier conditions.

As yet unrecognized needs related to carnivory could explain why *Catopsis berteroniana* presents the greatest challenge of all the Florida bromeliads. Perhaps unmet requirements related to prey dependence cause this plant to decline in culture despite receiving what constitutes appropriate care for its less fastidious relatives.

All of the Florida bromeliads ripen tiny seeds equipped with tufts of adhesive hairs (**Figure A, CP 16**) that secure them to bark until favorable conditions, meaning frequent wetting followed by rapid surface drying, induce germination. Growers can usually achieve yields exceeding those in nature using pieces of plastic screen or other similarly rough-textured, nonabsorbent materials as substrates.

Begin by using a spray of water to plaster a clump of seeds against a suitable medium. Then suspend the whole affair in a location exposed to gently circulating air. Water daily and germination should be evident within a few weeks. Year-old seedlings of the Florida natives won't exceed 3–10 millimeters in length, and don't expect to see flowers for at least 3–5 more years after that even under the most favorable growing conditions.

Should quicker results be desired, begin with mature plants and divide them into their component ramets. Then check to see which ones are likely to survive. Pups are worth nurturing when they possess leaves at least half as long as those borne by their parents. Firm brown, rather than soft white, tissue located at the separation point is an even more promising sign. If roots are beginning to develop, success is almost assured. Discard any post-flowering shoots that lack healthy-looking green leaves.

IDENTIFICATION KEYS

Identification of Florida's bromeliads to genus is easy if flowers or fruits are present. You can test your skill by using the following dichotomous keys. Dichotomous keys consist of series of paired, alternative choices that provide answers through a process of elimination. Each couplet contrasts one or more plant character states. Each pair of choices either leads to an identification of the unknown plant or to the next numbered, and usually indented, couplet, where the process is repeated. Once you come up with an answer, check it against the more detailed descriptions of the Florida species in the following section.

1a. Flowers borne on the inflorescence or its branches in spiral fashion. . . . 2

1b. Flowers borne solitary or in two ranks on the inflorescence axis or its branches . . . . ***Tillandsia***

2a. Floral bracts broad and conspicuous and mostly obscuring the inflorescence axis, flowers densely arranged . . . . ***Guzmania***

2b. Floral bracts small and inconspicuous, not obscuring the inflorescence axis, flowers laxly arranged . . . . ***Catopsis***

You can identify the species and named hybrids with the following three keys:

Tillandsia

1a. Leaves distichous: inflorescence 1- to 2-flowered.

2a. Stems elongate, plants forming pendant festoons; inflorescence reduced to a single, apparently sessile flower; petals yellow-green. . . . . 1. ***T. usneoides***

2b. Stems short, plants densely clustered to form spherical masses; inflorescence of (1–) 2 (–3) flowers, scape conspicuous; petals violet. . . . . 2. ***T. recurvata***

1b. Leaves polystichous: inflorescence 3- or more-flowered.

3a. Floral bracts imbricate, broad, covering all or most of rachis so that rachis is not visible at anthesis.

4a. Leaf sheaths conspicuously inflated; plants pseudobulbous.

5a. Leaf blades involute; scapes 1–15 cm; floral bracts densely lepidote.

6a. Inflorescence linear to linear-elliptic; base of floral bracts visible at anthesis. . . . . .3. ***T. paucifolia***

6b. Inflorescence broadly elliptic; base of the floral bracts not visible at anthesis. . . . . . . 4. ***T. pruinosa***

5b. Leaf blades channeled to involute; scapes 8–35 cm, floral bracts glabrous to inconspicuously lepidote near the apex only.

7a. Leaf blades and scape bracts contorted, reflexed. . . . . .5. ***T. balbisiana***

7b. Leaf blades and scape bracts erect to spreading. . . . . 6. ***T.*** x ***smalliana***

4b. Leaf sheaths flat or only slightly inflated, plants not pseudobulbous.

8a. Leaf blade linear-triangular to filiform.

9a. Leaf sheaths narrowly elliptic, 1.2–2.0 cm wide, slightly inflated. . . . . 7. ***T. simulata***

9b. Leaf sheaths broadly elliptic to triangular, 0.8–2.5 cm wide, flat.

10a. Leaves finely appressed lepidote, appearing green to reddish green; sheaths or scape bracts, especially the upper ones, narrowing abruptly into blades; floral bracts uniformly lepidote, green or tinged red; corolla lavender. . . . . .8. ***T. setacea***

10b. Leaves densely and coarsely appressed lepidote, appearing gray; sheaths of scape bracts narrowing gradually into blades; floral bracts uniformly red to rose, lepidote above, becoming sparsely lepidote toward the base; corolla violet.

11a. Leaves 15 to 30 in number, not over 5 mm wide at mid-length; inflorescence simple or with 1 to 5 lateral branches .9. ***T. bartramii***

11b. Leaves 20–50 in number, 5–8 mm wide at mid-length; inflorescence bipinnate with 2 to 10 lateral branches. . . . . .10. ***T.* x *floridana***

8b. Leaf blades obviously narrowly triangular, tapering evenly from base to apex.

12a. Leaves soft, brittle; inflorescence simple or laxly 2- to 3-branched, never digitate; floral bracts 6–9 mm wide. .11. ***T. variabilis***

12b. Leaves stiff, coriaceous; inflorescence densely digitate to laxly bipinnate with 3 to 15 branches; floral bracts 12–20 mm wide. . . . . 12. ***T. fasciculata***

3b. Floral bracts spreading and/or small, exposing most of the rachis at anthesis.

13a. Leaves 10 to 20 in number, spirally twisted and marked with bands; floral bracts 2.3–3.1 cm; inflorescence with 2 to 6 flowers per branch; corolla pink to dark rose. . . . . 13. ***T. flexuosa***

13b. Leaves greater than 20 in number, not twisted or banded; floral bracts 1.2–2.0 cm; inflorescence with 6 to 11 flowers per branch; corolla white. . . . . .14. ***T. utriculata***

Catopsis

1a. Inflorescence 3- to 10-flowered, pendant, corolla yellow, nocturnal. . . .
. . . . 1. ***C. nutans***

1b. Inflorescence 10- to 50-flowered, erect, corolla white, diurnal. . . . . . 2.

2a. Leaves yellow-green covered with white, chalk-like powder. . . . . . .
. . . . .2. ***C. berteroniana***

2b. Leaves bright green, not covered with white powder. . . . .
. . . . 3. ***C. floribunda***

Guzmania

1a. Leaf blades concolorous, green. . . . .
. . . . .1. ***G. monostachia*** var. ***monostachia***

1b. Leaf blades longitudinally green- and white-striped. . . . .
. . . . 2. ***G. monostachia*** var. ***variegata***

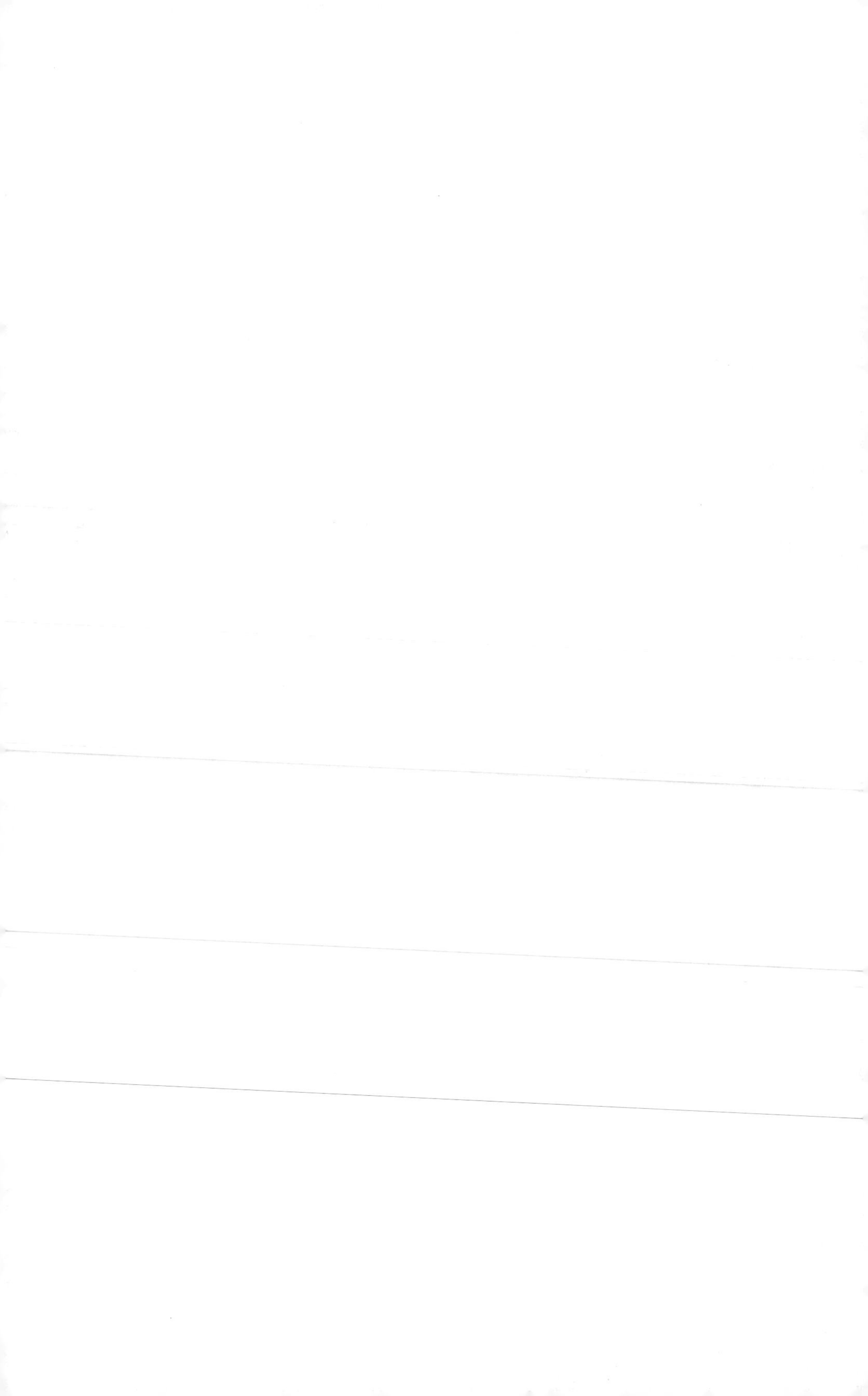

Guzmania monostachia, Fakahatchee Strand Preserve State Park (J. Staton photo)

CP 1 An immature specimen of *Tillandsia pruinosa* illustrating its characteristic dense cover of coarse foliar scales or trichomes (J. Staton photo)

CP 2 *Tillandsia fasciculata*, one of Florida's wild pines (B. Holst photo)

CP 3 *Tillandsia paucifolia* in fruit (B. Holst photo)

CP 4 Scanning electron micrograph of the surface of the leaf of *Tillandsia ionantha*, showing foliar trichomes and stomata (D. Benzing photo)

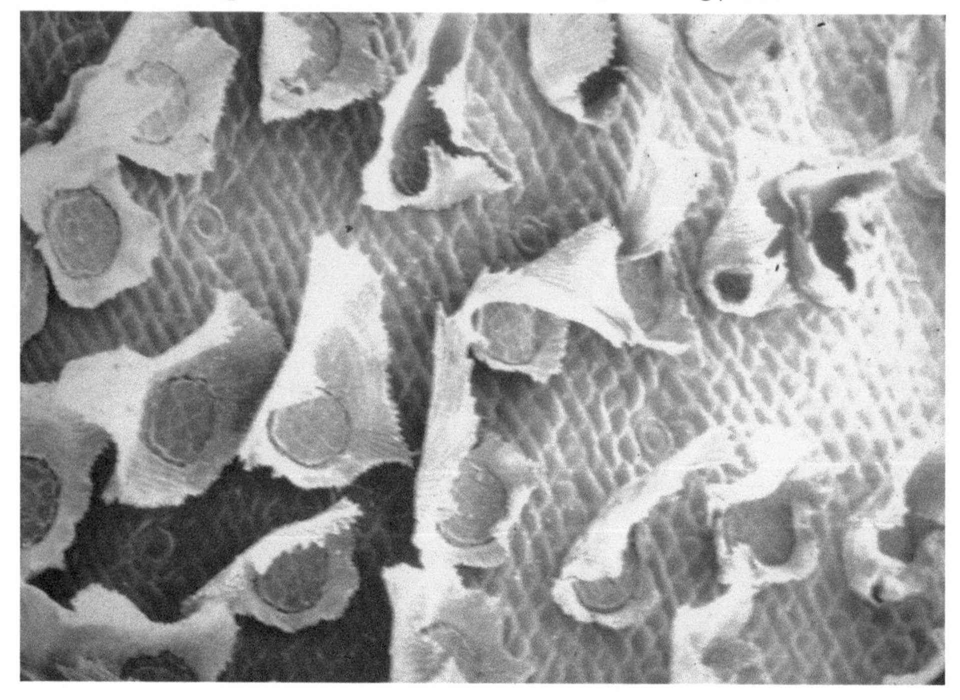

CP 5 Monocarpic *Tillandsia utriculata* in flower (B. Holst photo)

CP 6 Two ramets of *Tillandsia usneoides* in flower (B. Holst photo)

CP 8 *Tillandsia bartramii* (K. Marks photo)

CP 7 *Catopsis floribunda* dispersing seeds (B. Holst photo)

CP 9 *Tillandsia simulata* (K. Marks photo)

CP 10 *Guzmania monostachia* colony in the Fakahatchee swamp forest (K. Marks photo)

CP 11 *Tillandsia flexuosa* inflorescence illustrating a bird-pollinated flower (P. Nelson photo)

CP 12 Night-blooming *Catopsis nutans* in flower (J. Staton photo)

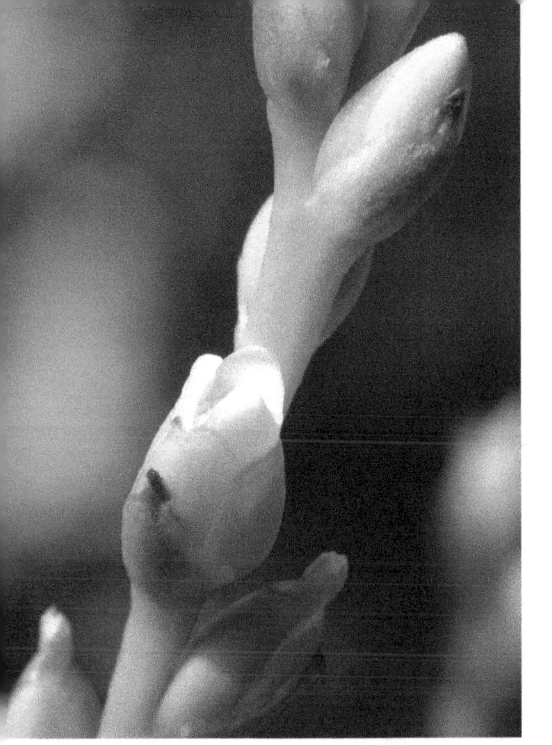

CP 13 Flower of *Catopsis floribunda* (K. Marks photo)

CP 14 Flower of *Tillandsia utriculata* (B. Holst photo)

CP 16 Seeds of *Tillandsia recurvata* (D. Benzing photo)

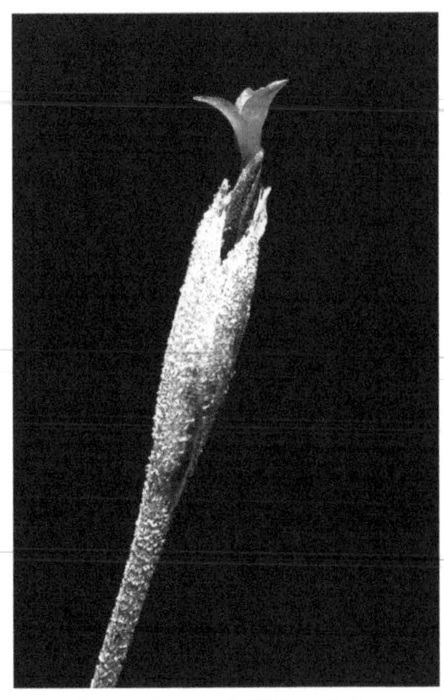

CP 15 Flower of *Tillandsia recurvata* (B. Holst photo)

CP 19 *Tillandsia utriculata* grown in bright sunlight
(D. Benzing photo)

CP 17 *Tillandsia recurvata* growing on a telephone cable
(B. Holst photo)

CP 18 *Tillandsia balbisiana*
(B. Holst photo)

CP 20 *Tillandsia utriculata* grown in shade (D. Benzing photo)

CP 21 A phytotelm-type bromeliad partially dissected to reveal decayed, decomposed litter in its leaf bases (D. Benzing photo)

CP 22 The disarticulated foliage of an immature *Tillandsia utriculata* specimen following the destruction of its stem by larvae of the Mexican weevil *Metamasius callizona* (B. Holst photo)

CP 23 *Tillandsia setacea* (K. Marks photo)

CP 24 *Tillandsia fasciculata* colony (B. Holst photo)

CP 25 *Tillandsia flexuosa* (B. Holst photo)

CP 26 *Tillandsia simulata,* cultivated on driftwood (L. Grashoff photo)

CP 27 *Catopsis berteroniana*: Note the presence of the loose waxy deposits on the leaf bases. (K. Marks photo)

CP 31 *Tillandsia recurvata*
(B. Holst photo)

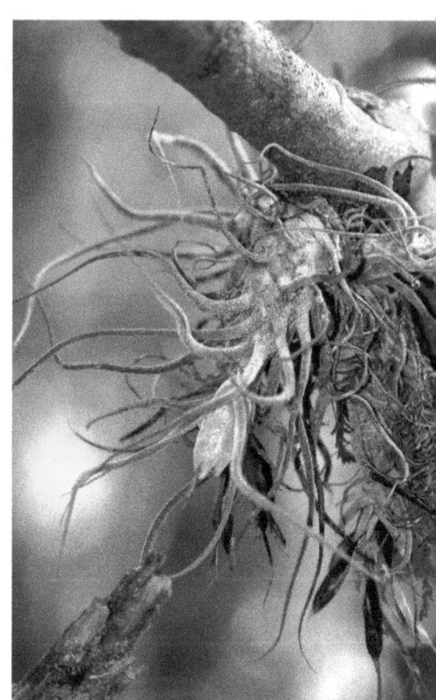

CP 30 *Tillandsia pruinosa*
(K. Marks photo)

CP 29 *Guzmania monostachia*
(K. Marks photo)

CP 28
Catopsis nutans
(J. Staton photo)

CP 32 A tree heavily laden with *Tillandsia setacea* (B. Holst photo)

CP 33 *Tillandsia setacea* flower (B. Holst photo)

CP 34 Citrus host in poor condition heavily laden with *Tillandsia usneoides* (L. Grashoff photo)

CP 36 *Tillandsia variabilis* inflorescence (B. Holst photo)

CP 35 *Tillandsia variabilis* (B. Holst photo)

CP 37 *Tillandsia x floridana* (K. Marks photo)

CP 38 *Tillandsia x smalliana* (K. Marks photo)

DESCRIPTIONS OF THE SPECIES AND HYBRIDS

The following descriptions of the bromeliad species—and where applicable their component subspecies, varieties, and forms, as well as the two hybrids that are also Florida natives—emphasize conspicuous features. Also provided are range maps that indicate by county where each taxon occurs within the state. Native habitats are described in terms of light conditions and humidity, along with any special plant features that represent adaptations for epiphytism. Each description also includes notes on natural history, such as when flowering occurs and any noteworthy relationships with animals.

CATOPSIS BERTERONIANA

(SCHULTES F.) MEZ

(CP 27)

Monogr. Phan. 9: 621. 1896.

This wide-ranging bromeliad flowers from 0.4–1.3 m tall, upright, vase-shaped, water-impounding shoots comprised of yellowish foliage bearing a powdery wax that renders the lower blades and bases chalky white. Leaves are broadly elliptic below, with the narrower, 2–4 cm wide blades tapering to points. Widely scattered, trichomes topped by narrow, flat caps cover both surfaces.

The upright inflorescence of *Catopsis berteroniana* features overlapping, leaflike bracts below; the upper flower-bearing portion remains simple or branches pinnately into 2 to 8 subdivisions, each of which supports up to 50 day-open flowers equipped with broadly ovate, 6–8 mm long yellow-green sepals barely exceeded in length by the white petals. The six stamens and single pistil remain largely concealed within the slightly spreading corolla tube. The resulting 1.2–1.4 cm long, ovoid, yellow to tan capsules split along three sides to release seeds bearing tufts of kinky brownish hairs at both ends.

High light–demanding *Catopsis berteroniana* flowers in fall and winter, and mostly from perches located among the uppermost branches and twigs of its diverse kinds of hosts. Tropical hardwood hammocks, open cypress woodlands, and buttonwood-mangrove forests support the densest of the surviving populations.

Observations and experiments conducted on Florida natives indicate botanical carnivory, which is to say that *C. berteroniana* feeds primarily on nutrients obtained from captured animals

(Frank and O'Meara, 1984; Benzing, 2000). Anchored high up on its hosts, it intercepts less litter than the more typical tank-bearing members of Bromeliaceae that depend on decaying plant material as their major source of mineral nutrients.

Prey capture by *Catopsis berteroniana* is passive, predicated on optical deception and tank confinement. Sunlight reflecting off the thickly wax-coated foliage tends to hide this epiphyte in plain sight from insects that routinely orient toward skylight as a way to avoid obstructions as they fly through forest canopies. Following collision with an unseen shoot, the hapless insect tumbles into one of its water-filled reservoirs where the insect drowns, unable to scale the adjacent, vertical leaf surfaces lubricated with the same wax responsible for its navigational error. Decomposition follows, mediated by tank-dwelling microbes that release soluble nutritive byproducts for use by the plant. The inability to secrete digestive enzymes on its own, or lure and actively trap prey, obliges this unusual bromeliad's designation as a proto-carnivore rather than a full-blown carnivorous plant such as the Venus flytrap.

Catopsis berteroniana appears to operate much the same way throughout its wide geographic range. Also worth noting is the fact that only one other bromeliad is demonstrably prey-dependent, and it belongs to the predominantly terrestrial genus *Brocchinia* in subfamily Pitcairnioideae.

CATOPSIS FLORIBUNDA

L. B. SMITH

(CPs 7, 13)

Cont. Gray Herb. 117: 5. 1937.

Shoots of this tank-equipped bromeliad attain heights of up to 60 cm when measurements include the erect inflorescence. Its numerous bright green leaves bearing scattered, barely visible trichomes are 20–40 cm long and equipped with broad, elliptic bases that taper to blades 1–3 cm wide. Blades borne by the outermost (oldest) leaves of mature shoots spread laterally to enhance the interception of precipitation and litter. Missing is the powdery wax that distinguishes *C. berteroniana,* and to a lesser extent *C. nutans.* If present here it would likely obscure the characteristic brownish hue of the leaf bases that identity all but the youngest seedlings. Contrary to *C. berteroniana,* the tanks of *C. floribunda* usually contain much decomposing plant matter and little, if any, remains of flying insects.

The upright inflorescences of *C. floribunda* emerge from fall to spring, after which they divide into two orders (meaning they are bipinnate) to yield 15 to 50 lateral branches. The main axis bears green, 4–5 mm long, ovate and somewhat spreading, overlapping floral bracts on the usually drooping subdivisions. Its many diurnal flowers display 6–8 mm long, white, slightly diverging petals above similarly elliptic 5–6 mm long yellow-green sepals (**CP 13**). Both the stamens and the stigma remain within the corolla tube. Seeds bearing tufts of brown hairs at both ends develop within stout capsules that split open on three sides. Fruiting is consistent enough to suggest sexual self-compatibility and perhaps a relaxed requirement for pollinators.

Florida *Catopsis floribunda* occupies humid, relatively dark forest understory habitats, most commonly where broad-leaved trees dominate. Favored sites include dense hammocks and sloughs; pond apple, red maple, Carolina ash, bald cypress, and oaks rank among the most heavily used hosts. Similar kinds of habitats support this species in parts of Mexico, Central America, and the Caribbean.

Restriction to Florida's southernmost counties and preference for shade accord with high vulnerabilities to frost and forest disturbance, respectively. Considerable appeal to browsing deer further threatens this species, as does its attractiveness to lubber grasshoppers that during major outbreaks severely graze the foliage. Quite possibly, water tables restored to former levels and protection will encourage some of Florida's much-reduced populations to rebound.

CATOPSIS NUTANS
(SWARTZ) GRISEBACH

(CPs 12, 28)

Fl. Brit. W. Indies 599. 1864.

Even long-established adults of this bromeliad consist of just one to several urn-shaped shoots, each comprised of relatively few spirally arranged, tightly overlapped, smooth, deep green, 8–15 cm long and 1–3 cm wide leaves with pointed tips. A layer of chalky wax often marks the lower portions of the outmost foliage, but less so than for *Catopsis berteroniana.* Like the other bromeliads with impounding shoots, this one grows upright, with colonies sometimes resembling rows of candles lined up along the upper sides of small branches. *Catopsis nutans* is by far the smallest of the three members of its genus present in Florida.

Three to 10 flowers associated with overlapped, leaflike, 1.4–1.5 cm long, elliptically shaped bracts attach directly to a pendant spike-type or sparingly branched inflorescence. Flowers feature green and leathery, 1.0–1.2 cm long elliptic sepals from which yellow, 2.0–2.4 cm long petals spread open at night. The resulting flared corolla and accompanying sweet fragrance attract nocturnal flyers, most likely moths with mouthparts capable of reaching sex organs and nectar positioned deep within the corolla tube. The ellipsoid, three-part, capsular fruits release seeds extended at both ends into tufts of kinky brown hairs.

Catopsis nutans is considered rare and endangered in Florida, perhaps even extinguishable by a single powerful hurricane (Ward, 1979). Its presence in the state consists of no more than a handful of small populations scattered across a few square miles or so of the Fakahatchee Slough in Collier County. Requirement for dark, understory habitats further heightens vulnerability to extirpation. Lacking the dense layers of reflective trichomes that protect many of Florida's other bromeliads from overexposure means that survivors probably stand little chance of rebuilding viable populations in severely wind-damaged forests.

Catopsis nutans flowers from fall through spring with seed dispersal timed to coincide with the beginning of the following winter dry season. High rates of fruit set indicate that Florida's populations can self-pollinate, perhaps unassisted by pollinators.

Catopsis nutans is easily distinguished from its two close Florida relatives. It alone bears a nodding pendant inflorescence—a feature recognized in the specific epithet "nutans." Individuals of the same species in Mexico and Central America display "imperfect" flowers, meaning that either functional stamens or pistils are present, but not both. Moreover, the male flowers are smaller than their female counterparts and borne on branched inflorescences rather than simple spikes. Florida's colonization by this bromeliad might never have occurred had the source population been entirely dioecious, in other words, comprised exclusively of individuals that produce either male or female flowers.

GUZMANIA MONOSTACHIA

(LINNAEUS) RUSBY EX MEZ.

(CPs 10, 29)

Monogr. Phan. 9: 905. 1896.

Including its erect inflorescence, the relatively squat, spreading rosulate shoots of *Guzmania monostachia* grow to 40 cm tall. Mature specimens consist of numerous, glossy, arching 30–40 cm long, thin, spirally arranged leaves with approximately 2 cm wide blades. Brown, sheathing leaf bases that form the same number of small impoundments usually hold abundant litter.

Closely inserted, overlapped floral bracts up to 2.5 cm long cover the cylindrical, many-flowered, undivided inflorescence. Those attached near its base feature dark green or reddish vertical stripes against a pale green background; closer to the apex the same appendages mostly lack associated flowers, but color up to pale salmon as the inflorescence matures.

Odorless, day-open flowers feature 1.4–1.6 cm long green sepals fused together at their bases. Slender, waxy, white petals 2.2–2.5 cm in length form a tubular corolla that encloses the shorter stamens and stigma. Fruits are ellipsoid, three-parted capsules 2–3 cm long that become brown when ripe. A parachute-like, hairy appendage attached at one end enhances airworthiness.

Being frost-sensitive and shade-demanding, judging by its extreme south Florida range and preferred anchorage low in the crowns of densely foliated trees, suggests a conservation status comparable to that of endangered *Catopsis nutans* and *C. floribunda.* This isn't so, however, because populations usually include hundreds of mature individuals and even higher densities of juveniles (**CP 10**). High germination and fruit set rates and parents that regularly generate multiple offshoots account for this impressive fecundity. Pond apple, Carolina ash, red maple, and some additional trees native to wet hammocks support the heaviest infestations.

Florida *Guzmania monostachia* neither looks like nor behaves the same way it does farther south through habitats that range into South America. Plants at these deeper tropical locations typically grow in fuller sunlight and display more colorful inflorescences whose flowers less regularly produce capsules.

Plants examined in the field in Trinidad shifted between C_3 and CAM photosynthesis as wet seasons alternated with dry ones (Benzing, 2000). Denied irrigation long enough during experiments, individuals from the same population exhibited similar behavior, including speedy recovery. After only a few hours following tank refilling, plants that had been stressed enough to begin absorbing carbon dioxide at night switched back to the more productive, but water-wasteful pattern of opening their stomata during the day (**Figure B**).

The fact that Florida *Guzmania monostachia* colors up less, and is more fruitful than its relatives farther south, may say something about how it managed to colonize the state. Likewise, its equally uncharacteristic shade tolerance provides a clue to how this species has come to tolerate Florida's relatively cold winters.

Long-distance colonizations by plants usually involve infrequent, long-distance seed dispersals. How likely a plant arriving as a seed will initiate a new population is influenced by many factors, not the least of which is the type of breeding system it possesses. To single-handedly establish a colony requires self-compatibility, yet *Guzmania monostachia* seems to be largely self-incompatible beyond Florida. It's also generally a sun-loving species, and physiologically quite a versatile one, judging by those performances recorded in Trinidad. Being able to shift between two modes of food manufacture imparts considerable drought tolerance and also heightens capacity to avoid photo-injury, a second benefit imparted by CAM photosynthesis.

If a single seed was responsible for establishing what we see of *Guzmania monostachia* in Florida today, then it must have germinated to yield one of the sporadic, self-fertile individuals that sometimes occur in generally sexually self-incompatible populations. It must also have been exceptionally shade tolerant for its kind, or it possessed a genotype amenable to natural selection for shade tolerance.

Frost sensitivity may explain why *G. monostachia* mostly grows deep within dense-canopied swamp forests and only in extreme south Florida. Doing so allows it to take advantage of the thermal insulation that dense vegetation provides. If this analysis is correct, then either the immigrant or immigrants were uncharacteristically shade tolerant for their species, or this capacity evolved in the population as repeated freezes eliminated its more sun-demanding members.

The dull inflorescence that accompanies self-compatibility assures that most growers prefer genotypes other than those from Florida except for *Guzmania monostachia* var. *variegata.* True to its name, representatives of this somewhat dubious variety, unlike the more widespread one (*G. monostachia* var. *monostachia*), bear foliage marked with conspicuous, lengthwise, chlorophyll-deficient stripes. The fact that variegated and concolorous (uniformly green) individuals grow side by side suggests that either viral infections or random matings that bring together defective pigment synthesis genes account for the atypical leaf coloration.

TILLANDSIA BALBISIANA

SCHULTES f.

(CP 18)

Roemer & Schultes. Syst. 7(2): 1212. 1830.

No other member of genus *Tillandsia* native to Florida features foliage as strongly reflexed or as characteristically twisted as that displayed by *T. balbisiana.* Equally useful for identification is the way that these same leathery leaves flare at the base and overlap to form a bulblike arrangement. Equipped with wiry, nonabsorptive roots but no tanks, this bromeliad operates in the atmospheric mode, meaning that it obtains nutrients and water exclusively from dust and fluids that contact its absorptive foliage.

A dense covering of light-reflecting and moisture- and nutrient-absorbing trichomes renders the entire 30- to 35-leafed shoot of *T. balbisiana* silvery gray except at its base, which is suffused with soft brown. Should foliage be exposed to intense sunlight, it colors up to a diffuse red to help shield the underlying green tissues. Individuals raised in shade brighten up only when pollinators are needed, and coloring tends to be weaker and mostly confined to the inflorescence.

Inflorescences grow upright, are quite tall (8–30 cm) for the size of the supporting shoot, and usually branch bipinnately into 2 to 10 erect subdivisions. Each of these subdivisions is 2–10 cm long, flattened, and equipped with 5 to 12 flowers associated with overlapped, 1.5–2.0 cm long, bright red to green floral bracts. Leathery, keeled, and sharply tipped, each one bears a single flower in its axil. Spreading to recurved in shape, the more leaflike scape bracts arise along the inflorescence axis below its fertile region.

Flowers that open during the day feature smooth, 1.5–2.0 cm

long green sepals inserted below a tubular corolla consisting of 3 strap-shaped, violet petals up to 3.5 cm long. Both the stamens and stigma extend beyond the corolla, the former endowing it with a bright yellow tip while the pollen is being presented for pollinator pickup. Mature capsules are either green or brown, cylindrical, and up to 4 cm long. Seeds bear parachute-like tufts of hairs at one end to promote dispersal by wind. Floral morphology, daytime receptivity, and the production of nectar but no fragrance, identify hummingbirds as the targeted pollinators.

Tillandsia balbisiana prefers moist to moderately dry conditions on hosts ranging from bald cypress to coastal strand shrubs throughout the southern half of peninsular Florida. Although much reduced in numbers by over-collection and habitat destruction, dense populations persist in some protected areas.

Flowering occurs from spring into summer. Northernmost colonies in Florida fail to develop the red coloration associated with reproduction farther south. High rates of fruit set further suggest that these outliers produce seeds without assistance, i.e., are at least self-fertile and probably capable of spontaneous self-pollination.

Tillandsia balbisiana engages ants in two ways that potentially favor both the plant and its insect attendants. Individuals encountered in Mesoamerica often harbor colonies of aggressive ants in the dry cavities within the onion-shaped bases of their shoots, presumably for defense against herbivores; plant feeding is a second possibility. The occasional Florida resident also houses ants in its shoots, but not the same species that attack when large animals disturb their nest sites in places like southern Mexico.

The other relationship between this bromeliad and ants is less intimate and protective only, and then just during the flowering season. Ants are attracted in this second case by nectar secreted from glands located on the floral bracts rather than deep within the flowers. This temporary, sugar-rich bounty can be substantially supplemented for bodyguards swift enough to capture herbivores or more casual visitors unaware that they have entered patrolled territory. Following fruit set, the extrafloral nectaries dry up, forcing any remaining ants to seek new feeding stations on younger inflorescences.

TILLANDSIA BARTRAMII

ELLIOTT

(CP 8)

Bot. S. Carol. & Georgia 1: 378. 1817.
Synonym: *Tillandsia myriophylla* Small,
Man. S. E. Fl. 270, 1503

This cluster-forming bromeliad flowers when 20–45 cm high. Reproducing shoots bear 15 to 30 erect to spreading leaves characterized by triangular, rust-colored bases incapable of impounding significant quantities of moisture or litter. The slender (2–5 mm wide) and leathery blades are involute (in-rolled upward) and gray and scurfy owing to the presence of dense layers of light-reflecting scales. Shoot orientation often deviates from the upright posture characteristic of bromeliads with better-developed, leafy tanks.

The 8–15 cm long inflorescence of *Tillandsia bartramii* is erect and either simple or divided, the 2 to 5 branches being flattened, narrow, and pointed at their apices. Each subdivision features 5 to 20 spirally arranged flowers inserted in the axils of an equal number of keeled, acutely pointed 1.4–1.7 cm long floral bracts. Overlap is sufficient to obscure the supporting rachis. Suffused with red to rose pigments during flowering, they fade back to green as the fruits mature.

Tillandsia bartramii, along with most of the other members of its genus in Florida, produces odorless, diurnal flowers equipped with tubular corollas from which the stamens and stigma protrude (**Figure A**). Birds are the likely pollinators, although most individuals set abundant fruits, suggesting spontaneous self-pollination. The sepals are elliptic and keeled, up to 1.5 cm long, scurfy like the floral bracts, thin, and leathery. The petals are 3.0–4.5 cm long and violet, and the resulting fruit three-parted, cylindrical, 2.5–3.0 cm long, and filled with seeds that bear parachute-shaped tufts of hairs at one end (**Figure A**).

Tillandsia bartramii grows on many kinds of hosts in humid woods and riverside swamps, where it flowers from March to July. Dense colonies often occupy the trunks of bald cypress, oaks, red cedar, and red maple. In the northern part of its range, plants tend to occur low on their hosts near water in wetlands, perhaps because frost precludes higher perches. Nevertheless, this species appears to be one of the more cold-hardy of the Florida bromeliads.

Tillandsia bartramii can be distinguished from close look-alike *T. simulata* by its tendency to grow in dense hemispherical clusters rather than as small erect clumps or single shoots. Additionally, the leaf sheaths of *T. bartramii* are triangular and obscure compared to their more elliptic and conspicuously inflated conditions in *T. simulata. Tillandsia setacea,* another potential source of confusion, is considerably greener than either of the former species, and its leaf blades are often tinged with red (**CP 23**). Plants virtually indistinguishable from Florida *T.*

bartramii occur in northeastern Mexico.

TILLANDSIA FASCICULATA

SWARTZ

(CPs 2, 24)

Prod. Veg. Ind. Occ. 56. 1788.

Common name: Cardinal air plant

Like all but one of the state's other bromeliads, Florida's population of *Tillandsia fasciculata* marks the northernmost boundary of the range of a fundamentally more tropical species. In this case, however, outlier status is not associated with the diminished colors of flowers and floral bracts that so often signal sexual self-compatibility. Consult the treatments of *Catopsis nutans* (page 67) and *Guzmania monostachia* (page 69) for details concerning what breeding systems and floral pigmentation may reveal about the migratory histories of the Florida natives.

Flowering occurs when the shoot of *Tillandsia fasciculata,* including its inflorescence, is 40–65 cm tall. Typically, a compliment of 20 to 50 gray to gray-green, spirally arranged, spreading leaves, each with a dark, rust-colored, expanded base, impound modest amounts of water during the rainy season. The 2–3 cm wide blades, being more or less upright, stiff, and channeled, also intercept considerable litter. A confluent layer of foliar scales allows the entire shoot to reflect light when its surfaces are dry and to absorb moisture and nutrients while wet.

Inflorescences are erect to ascending and 10–35 cm tall. Progressively smaller, overlapping, erect to spreading leaflike bracts cover the lower, sterile part of the main axis (the scape) upward to its point of branching. Once-pinnate branching results in 3 to 15, 5–20 cm long, either erect or spreading divisions that individually bear 10 to 50 flowers. Inserted immediately below each one is a keeled, sharply pointed, overlapping, 1.2–4.8 cm long, mostly naked floral bract. During flowering, much of the inflorescence colors up to rose, intense red, yellow-green, or a combination of these colors, depending on the variety and the plant's exposure to sunlight. Red to orange pigments may also suffuse the foliage.

Flowering occurs from mid-spring into summer. The lanceolate sepals taper to acute tips, are about 4 cm long, keeled above, and naked to slightly scaly. The strap-shaped, 5–6 cm long petals align to form a tubular, violet, blue, or, rarely, white corolla beyond which the stamens and stigma extend several millimeters. As with most of the other Florida tillandsias, this one is equipped to attract avian pollinators. Seeds with a parachute-like hairy appendage at one end develop within approximately 4 cm long cylindrical, three-parted capsules.

Only *Tillandsia utriculata* (**CP 5**) exceeds *Tillandsia fasciculata* for overall size, except in especially favorable habitats—for example, through much of the Big Cypress Preserve, where long-establish specimens of *T. fasciculata* var. *densispica* consist of more than a dozen crowded shoots (**CP 24**).

Many kinds of trees and shrubs host this epiphyte, and in diverse types of habitats. Frequent occurrences in upland sites on trees with thinly foliated crowns indicate considerable drought-hardiness, although the densest populations develop under more humid conditions. Moderate sensitivity to frost by Florida bromeliad standards may explain why *Tillandsia fasciculata* more faithfully associates with wetlands in the northern part of its range. Greater exposure to fire at drier sites may further account for this distribution.

Field-collected *Tillandsia fasciculata* more often ends up in gardens and managed landscapes around the state than any of the other native bromeliads. Transplants color up in their new surroundings much as in the wild, the bright, candlelike inflorescence and epiphytic habit more than justifying the label cardinal air plant. On the down side, *Tillandsia fasciculata* shoots became large enough to attract abundant egg laying by the destructive Mexican weevil, *Metamasius callizona.* How seriously this recently introduced herbivore threatens *T. fasciculata* and the other large-bodied Florida bromeliads is discussed on pages 45–46.

The formal recognition of three varieties and one form just within Florida's *Tillandsia fasciculata* population bears witness to its exceptional polymorphism. Additional named segregates occur elsewhere across its broad, mostly Mesoamerican range. *Tillandsia fasciculata* also has several structurally similar, close relatives outside Florida, suggesting multiple divergences from a recent and prolific common evolutionary ancestor.

Note that the major segregates of *Tillandsia fasciculata* constitute varieties, whereas those of *T. utriculata* are designated subspecies. Readers can consider these two taxonomic ranks biologically equivalent.

The following four are named segregates of *Tillandsia fasciculata*:

TILLANDSIA FASCICULATA VAR. *CLAVISPICA* MEZ

D.C. Monogr. Phan. 9: 683. 1896.
Synonym: *Tillandsia bracteata* Chapman,
Fl. South. U.S. 471. 1896.

The inflorescence of variety *clavispica* divides more laxly than that of more widespread variety *densispica.* Also distinctive is the greater length of the sterile bases of the lateral branches compared to their apical, 10 cm or shorter portions, where the overlapping, larger, flower-bearing floral bracts insert. Restricted to south Florida, its members tend to grow in bright light on numerous kinds of trees, where they readily hybridize with *T. fasciculata* var. *densispica.* Inter-breeding is so extensive that mixed populations contain few individuals that comfortably match the formal descriptions of either of the two parents. Many of the putative hybrids exhibit pale to pastel bract and petal coloration.

TILLANDSIA FASCICULATA VAR. *DENSISPICA* MEZ

Monogr. Phan. 9: 983. 1896.
Synonym: *Tillandsia hystricina* Small,
Man. S.E. Fl. 271. 1933.

The inflorescence branches of this second variety cluster—each one featuring a short, sterile basal portion topped by a terminal, fertile extension that rarely exceeds 10 cm. Floral bracts up to 3 cm long range from dull to bright red at flowering. This most common and wide-ranging of the varieties in central and south Florida utilizes diverse kinds of hosts. It is also the most popular selection for cultivation.

TILLANDSIA FASCICULATA* VAR. *FASCICULATA SWARTZ

This third Florida variety is known only from a single collection made in southern Miami-Dade County, and its current status in the state is unknown (Luther, 1993). Variety *fasciculata* could be considered aberrant for the species were it not so similar to plants widely distributed throughout the Caribbean and Central America. Florida variety *fasciculata* does differ from its putative, more southerly counterparts by its yellow-green, rather than red, 3–5 cm long floral bracts.

TILLANDSIA FASCICULATA* VAR. *DENSISPICA* FORMA *ALBA M. B. FOSTER

Bromel. Soc. Bull. 3: 29. 1953.

Little, if anything, beyond bracts and petals that lack anthocycanin pigments differentiate this plant from the violet-flowered, typical version of the same variety. Only one of the genes that mediate the biosynthetic pathway responsible for the production of the bromeliad's red to purple pigments need be defective to account for such pale petals or bracts. Whether this single distinguishing feature justifies assigning a formal name is arguable. Quite plausibly, these plants represent crosses between varieties *densispica* and *clavispica;* hybrids may be less colorful than one or both of their parents.

TILLANDSIA FLEXUOSA

SWARTZ

(CPs 11, 25)

Proc. Veg. Ind. Occ. 56. 1788.
Synonym: *Tillandsia aloifolia* Hooker, Exot. Fla.: pl. 205. 1827

Tillandsia flexuosa flowers from shoots up to a full meter tall when measurements include the outsized, spreading inflorescence. Ten to 20 gray to reddish, spirally arranged, 30–40 cm long, distinctly silver-banded leaves overlap to form the characteristic bulblike base that appears better suited to shelter small animals than to impound water or litter. Narrowly triangular blades, which are 10–25 mm wide where they join the reddish brown, ovate to elliptic leaf bases, spread upward and away from the shoot.

The spiral twisting and faint horizontal bands that so definitively distinguish this bromeliad already have become evident by the time seedlings reach about a centimeter long. Because spent shoots deteriorate rapidly, the ratio of parents to daughter ramets approaches one. Even long-established specimens seldom display the clustering characteristic of so many of the other tillandsias, such as *T. fasciculata*.

The lower, sterile portion of the 15–40 cm tall, 5- to 40-flowered inflorescence bears spirally arranged, leaflike scape bracts that diminish in size upward to the point where the lowest of the usually 2 to 8 pinnately arranged branches arise. These subdivisions stand more or less erect, with their spreading bracts and flowers inserted far enough apart to leave much of the supporting axis exposed. Flowering may induce the 2.3–3.1 cm, prominently nerved, sparsely scaled floral bracts to develop a deep red color.

Day-opening flowers display gray-green, reddish-tinged, elliptic, 2–3 cm long, conspicuously nerved sepals. Four-centimeter-long, strap-shaped petals ranging from pink to dark rose form a tubular corolla that flares modestly at its summit. Somewhat longer stamens and stigma protrude beyond the petals. Three-part, cylindrical capsules up to 7.5 cm long release airborne seeds equipped with an umbrella-shaped, hairy appendage at one end. Flowering occurs during the summer.

Tillandsia flexuosa thrives in bright sun, and its penchant for coastal strand and cypress swamp habitats in extreme south Florida suggests pronounced frost sensitivity. Buttonwood and cypress, among many other kinds of trees and shrubs, provide its anchorages; formerly dense populations on dwarfed cypress along the northwest edge of the Big Cypress Swamp have largely disappeared, most likely owing to lowered water tables and consequent fire. Repeated severe freezes during the 1970s and 1980s increased the toll.

TILLANDSIA PAUCIFOLIA

BAKER

(CP 3, Figure A)

Gard. Chron. II. 10: 748. 1878.

Tillandsia paucifolia is a compact-bodied, densely trichome-covered bromeliad that flowers from shoots usually no more than 15 cm long. Five to 10 spirally arranged, gray leaves with broadly ovate bases overlap to form an elongate bulblike main body. The blades are inrolled and about the same length as the base. Shoots orient in every direction and do not impound moisture; instead, ants with brood often take up residence inside, gaining access to the leaf base chambers via holes that they or other creatures produce. Roots are wiry, often with just one or a few of them securing plants to their substrates.

Unlike the leafy portion of the shoot, the 1–6 cm long inflorescence grows upright, curving as needed to achieve this orientation. Only those of the most robust specimens often don't grow upright, and they may branch sparingly. The inflorescence bracts closely overlap beginning as leaflike appendages below and becoming increasingly flattened upward, especially those that subtend the 1 to 15 flowers. Bracts inserted just below the fertile part of the inflorescence, and those associated with the flowers (the floral bracts), are 2–3 cm long and broadly acute to obtuse. They become deep pink during flowering. Lesser accumulations of the same pigments more delicately color the foliage of individuals exposed to intense sunlight.

Bird-adapted flowers open during the day from early spring into summer to display up to 4 cm long tubular, lavender-blue corollas topped by six exerted stamens and the stigma. The sepals are lanceolate, up to 2 cm long, acute, thin, keeled, and sparsely scaly. Fruits mature into cylindrical, up to 4 cm long, three-parted, brown capsules that contain around 150 seeds equipped with umbrella-like tufts of hairs at one end.

Tillandsia paucifolia occurs most abundantly on dwarfed cypress and other hosts similarly characterized by thinly foliated canopies. Many dozens of adults infesting the crowns of single trees used to be common in the Big Cypress Swamp. Specimens growing where bark is more heavily shaded, while less abundant, are often more massive and fruitful.

Florida's *T. paucifolia* has been misidentified as *T. bulbosa* and *T. circinnata.* The first label applies to another species that occurs on the nearby Bahamas and through much of the rest of Mesoamerica. The second one appears in some references that deal with native Florida flora and certain cultivated plants and represents a synonym of a Mexican and Central American species.

Field and laboratory studies conducted on this wide-ranging species have revealed much about the ecophysiology and natural history of the atmospheric bromeliads (Benzing, 1981, 2000).

TILLANDSIA PRUINOSA

SWARTZ

(CPs 1, 30)

Fl. Ind. Occ. 1: 594. 1797.

Tillandsia pruinosa parallels *T. paucifolia* for small size, bulbous form, and the way shoots grow with no consistent orientation. Here as well, 5 to 10 spirally arranged leaves form the mature shoot, each one consisting of an expanded, reddish brown sheath that abruptly tapers into a relatively short, involute, 3–5 mm wide, undulating blade. A continuous layer of coarse-textured, highly reflective scales that clothes the entire shoot, including its floral bracts, both distinguish this species from *T. paucifolia* and account for its colloquial name, which is hoary air plant. The same applies for "pruinosa," a Latin term that alludes to frost.

Three to 12 flowers emerge from between two opposing ranks of overlapping, ovate and keeled, 2.0–2.2 cm long floral bracts inserted along a 1–3 cm long, flattened inflorescence that does not curve upward as is usual for *T. paucifolia*. Flowers open during the day, their blue-violet corollas accentuated by temporarily deep pink floral bracts and adjacent foliage suffused with the same pigments. The elliptic sepals are keeled, thin, and scale-free, as is the up to 3 cm long tubular corolla comprised of strap-shaped petals beyond which the stamens and stigma protrude. The resulting fruit is a cylindrical brown capsule, which being up to 3.5 cm long, seems oversized relative to the rest of the shoot. Like the rest of the tillandsias, the seeds possess an umbrella-like fluffy appendage at one end.

This infrequently seen bromeliad, like *T. paucifolia,* often hangs from its hosts by one or a few wiry roots. Despite its compact habit, considerable leaf succulence, and well-developed covering of light-reflecting scales, it favors shady understory habitats, mostly in humid locations in the Big Cypress and Fakahatchee swamps. Many kinds of trees host it, among the most extensively exploited being cypress, pond apple, red maple, oak, and Carolina ash.

Florida's *Tillandsia pruinosa* is restricted to the extreme southern part of the state where its populations consist of modest numbers of widely scattered individuals. Its status as the most recently discovered of the Florida bromeliads (1947) reflects the thinly stocked nature of its populations and their relegation to remote habitats.

TILLANDSIA RECURVATA

(LINNAEUS) LINNAEUS

(CPs 15–17, 31)

Sp. Pl. 2. 410. 1762.

Synonym: *Diaphoranthema recurvata* (Linnaeus) Beer, Bromel. 156. 1857

Common name: ball moss

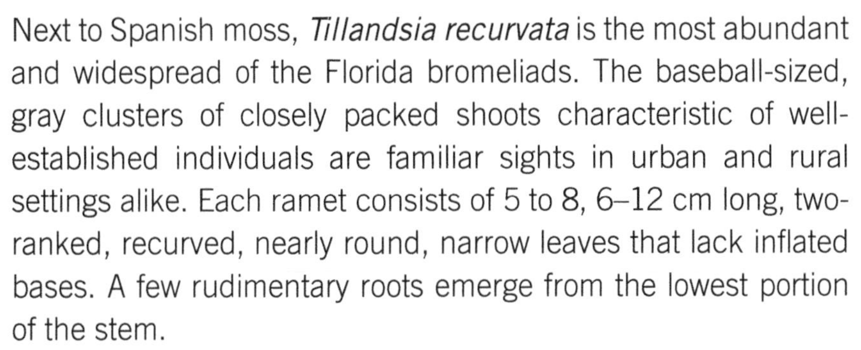

Next to Spanish moss, *Tillandsia recurvata* is the most abundant and widespread of the Florida bromeliads. The baseball-sized, gray clusters of closely packed shoots characteristic of well-established individuals are familiar sights in urban and rural settings alike. Each ramet consists of 5 to 8, 6–12 cm long, two-ranked, recurved, nearly round, narrow leaves that lack inflated bases. A few rudimentary roots emerge from the lowest portion of the stem.

Lacking both absorptive roots and tanks, *T. recurvata,* along with Spanish moss, epitomizes the atmospheric way of life. Its frequent occurrence on telephone wires (**CP 17**) and attachments to other inert substrates also reinforces the notion that the gray tillandsias possess the capabilities that their popular designation as air plants falsely implies. While no plant can sustain itself exclusively on air, ball moss and the atmospheric bromeliads in general are the consummate botanical scavengers of airborne nutrients. How they acquire moisture and nutrients and from where are explained under the section on page 27 titled "The leaf scale."

One or two flowers occupy the summit of the 2–5 cm long inflorescence, which is naked, except for a single leafy bract. The narrowly elliptic, green-tinged to purple floral bracts overlap to mostly obscure the short rachis. Lavender, 0.7–1 cm petals emerge above the somewhat shorter green sepals. The stamens and stigma remain hidden within a tubular corolla that spreads open ever so modestly at its apex. Diminutive flowers like these that also produce no nectar or fragrance signal long-established reliance on self-set seeds. Indeed, brown, up to 3 cm long, three-parted capsules containing seeds with the usual umbrella-shaped, hairy appendage consistently form on specimens maintained in pollinator-free spaces.

Tillandsia recurvata requires bright light, and it flowers during the summer. Its exceptional abundance, often on trees in conspicuous decline, suggests parasitism. In fact, this bromeliad may be particularly adept at stressing its hosts by absorbing nutrients that naturally leach from its foliage as described in the discussion of the phenomenon called nutritional piracy in the section titled "Do bromeliads ever harm their hosts?" on page 36.

It's no coincidence that ball moss is the closest rival to *T. usneoides* for geographic range. Both species belong to the same small subset of closely related tillandsias whose ancestors resided in Central Andean South America, where the group continues to be most species-rich. Ball moss just reaches the southeastern corner of Georgia, whereas even more successful Spanish moss, probably owing to superior winter-hardiness and propensity to reproduce by fragmentation in addition to seeds, extends northward as far as maritime Virginia.

A second population of *T. recurvata* anchored in Central America continues north through Mexico and east into Louisiana. New colonies probably initiated by immigrants imported on infested landscape stock have appeared even farther eastward during just the past several decades.

TILLANDSIA SETACEA

SWARTZ

(Figure A, CPs 23, 32, 33)

Fl. Ind. Occ. 1: 593. 1797.

Synonyms*: Tillandsia caespitosa* LeConte,
Ann. Lyc. Nat. Hist.
N.Y. 2: 131. 1826.
Tillandsia tenuifolia Linnaeus, Sp. Pl. 286. 1753 in part.

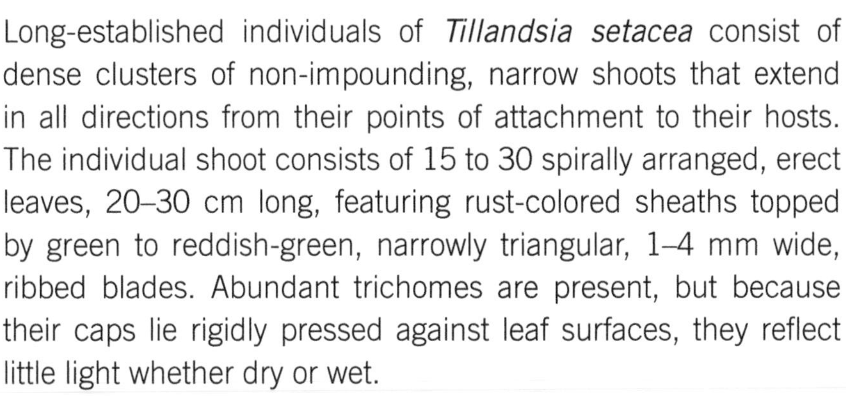

Long-established individuals of *Tillandsia setacea* consist of dense clusters of non-impounding, narrow shoots that extend in all directions from their points of attachment to their hosts. The individual shoot consists of 15 to 30 spirally arranged, erect leaves, 20–30 cm long, featuring rust-colored sheaths topped by green to reddish-green, narrowly triangular, 1–4 mm wide, ribbed blades. Abundant trichomes are present, but because their caps lie rigidly pressed against leaf surfaces, they reflect little light whether dry or wet.

The 3- to 15-flowered inflorescence branches to produce a tight array of 2 to 5 flattened, erect 1–4 cm long branches, each of which bears leathery, 0.8–1.2 cm long, ovate, green or reddish, overlapping floral bracts with acute tips. Flowers open during the day to expose 1.8–2.2 cm long, lavender strap-shaped petals that form tubes through which the stamens and the stigma protrude (**Figure A, CP 33**). The sepals are elliptic, 0.8–1 cm long, and sparsely scaled. Fruits ripen into 2–3 cm long, cylindrical, three-parted capsules, which upon dehiscence release seeds equipped at one end with an umbrella-shaped, hairy appendage.

Tillandsia setacea flowers from spring to fall. Its most heavily

utilized hosts include Carolina ash, cypress, pond apple, and several kinds of oaks. Low branches in swamp forests support the most crowded colonies across a range that extends the full length of the peninsula (**CP 32**). *Tillandsia setacea* is also probably the most abundant Florida bromeliad after *T. usneoides* and *T. recurvata*. It, along with these other two tillandsias, may also head the list for frost tolerance, in this case responding to winter temperatures by becoming more intensely red.

Tillandsia setacea is easily distinguished from structurally similar *T. bartramii* and *T. simulata* (**CPs 8, 9**) by its much greener, narrower foliage. The name *Tillandsia tenuifolia* has been misapplied to *T. setacea* in some of the br

TILLANDSIA SIMULATA

SMALL

(CPs 9, 26)

Man. S. Fl. 270, 1503. 1933.

Tillandsia simulata flowers from shoots up to 40 cm tall bearing 15 to 30 spirally arranged, non–water-impounding leaves. The modestly inflated, brown bases of these organs continue upward as narrowly linear-triangular, 1–4 mm wide, involute, sharply pointed blades. A confluent layer of light-reflecting trichomes lends the shoot its decidedly gray color.

The inflorescence bears 5 to 30 diurnal flowers on as many as five, upright, spreading and flattened lateral branches. Its erect, broad and overlapping, 1.4–1.8 cm long, rose-colored floral bracts mostly obscure the subtending rachis. Scales like those on the foliage occur on the floral bracts, and to a lesser extent, also on the up to 1.6 cm long sepals. The tubular violet corolla is shorter than the protruding stamens and stigma. Seeds equipped on one end with an umbrella-shaped collection of hairs develop within approximately 3 cm long cylindrical capsules.

Tillandsia simulata flowers in the spring and utilizes a variety of kinds of hosts, especially southern red cedar, cypress, oak, and *Magnolia.* Favored habitats are swamps and moist hammocks where the most heavily exploited microsites feature strong light. *Tillandsia simulata* is Florida's only endemic bromeliad. Several close relatives share similar shoot architecture, and in Florida *Tillandsia simulata* is easily confused with *T. bartramii* (**CP 8**). Most distinctive is the leaf base, which is elliptic and inflated in the former species and triangular and flat in the latter.

TILLANDSIA USNEOIDES

(LINNAEUS) LINNAEUS

(CPs 6, 34)

Sp. Pl. ed. 2. 411. 1762.

Synonym: *Dendropogon usneoides* (Linnaeus) Rafinesque. Fl. Tellur. 4: 25. 1838.

Common name: Spanish moss

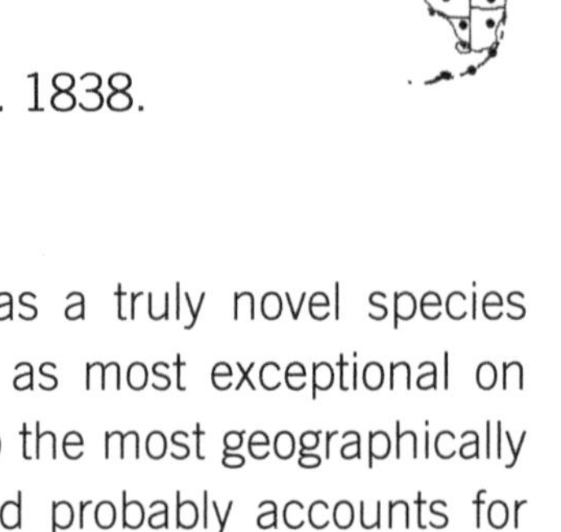

Spanish moss warrants special note as a truly novel species among the flowering plants, and also as most exceptional on several counts within its family. It's also the most geographically dispersed of all of the bromeliads, and probably accounts for more biomass than any of them. Only the duckweeds, a few additional aquatics in several other families, and a handful of species with largely subterranean or parasitic habits exceed it for anatomical reduction.

Adults are essentially rootless with the vegetative portion of the individual shoot or ramet amounting to just three, nearly tubular, densely scaled leaves inserted in two ranks on a short stem. Long-established adults consist of thick festoons comprised of up to 200 cm long, intertwined strings of serially attached ramets (**CP 34**).

Flowers borne singly on a short rachis barely extend above the sheathing base of the sub-adjacent leaf (**CP 6**). The only other organ present on the inflorescence is a single broad, erect green floral bract 4–5 mm long, which like the foliage, is densely covered with coarse, reflective scales. The sepals are naked, 6–7 mm long, and ovate with acute tips. Its three yellow-green, spoon-shaped petals spread enough to reveal the tops of the included stamens and stigma. A pronounced sweet fragrance emitted beginning late in the afternoon and continuing into the night indicate moth pollination. Flowers are followed by up to 2.5 cm long, cylindrical capsules that release 20 to 50 seeds, each with an umbrella-shaped, hairy appendage at one end.

Tillandsia usneoides owes its record geographic range to multiple factors. Among them are capacities to tolerate prolonged frost and exposures ranging from full sun to medium shade. Its tendency to fragment and its allure to nest-building birds aid dispersal. High rates of fruit set foster much sexual reproduction, the resulting seedlings anchoring on bark with tiny wiry roots. An unusually wide host range further contributes to success, as does the suitability of telephone wires and other inert supports as alternative substrates.

Surprising vigor for a plant entirely reliant on its shoots to absorb moisture and nutrients from precipitation and aerosols is yet another advantage enjoyed by this bromeliad. Why growth can be so rapid isn't obvious, although certain material economies attending its much-abbreviated body probably contribute. Further advantage accrues from the shoot's much-reduced surface-to-volume ratio, a condition that may promote photosynthesis by enhancing gas exchange and light harvest. On the other hand, it makes food using the CAM mechanism (see pages 21–25).

All of the atmospheric epiphytes likely benefit from anatomical modifications that allow them to allocate unusually large proportions of key resources like nitrogen to green rather than non-photosynthetic tissue. Virtually the entire shoot of *Tillandsia usneoides* makes food, and bearing absorbing trichomes to boot, it also takes up moisture and nutrients throughout. Nothing need be invested in roots, and indeed, nothing usually is after completing the seedling stage.

Also worth comment is the demonstrated value of Spanish moss and ball moss for monitoring air quality in Florida and several

Latin American nations. Both species accumulate heavy metals, among other wind-borne contaminants, to high concentrations. Spanish moss collected from trees near a Florida highway for one study contained abnormally high concentrations of lead (Martinez et al., 1971). A few years later, after tetraethyl lead had been eliminated from gasoline, specimens from the same sites had become chemically indistinguishable from those collected well removed from automotive traffic.

TILLANDSIA UTRICULATA

LINNAEUS

(CPs 5, 14, 19, 20)

Sp. Pl. 286. 1753.

Common name: giant wild pine

Faithful to its colloquial name, *Tillandsia utriculata* produces the largest shoots of any of Florida's bromeliads. It is also the only monocarpic (single-flowering) species among the state's 16 native species, meaning that it alone dies without producing offshoots, propagating exclusively by seeds. *Tillandsia utriculata* probably has to become this large to impound sufficient moisture and litter to produce enough seeds to sustain its populations. Rather than the hundreds to several thousand per inflorescence characteristic of species like *T. paucifolia* and even larger *T. fasciculata,* the number of seeds produced by a typical *T. utriculata* specimen exceeds 10,000.

Shoots grow up to two meters tall (**CP 5**), by which time they bear 20 to 75 spirally arranged, 0.3–1 m long, strap-shaped leaves with massively inflated (utriculate), rust-colored bases topped by broadly triangular blades. The confluent layers of flattened scales that cover these 2–7 cm wide, channeled blades allow the underlying pigments to show through. Stiff, upright, grayish-purple foliage is characteristic of specimens grown in bright light (**CP 19**); individuals anchored deeper in the canopy feature greener shoots comprised of softer, drooping leaves (**CP 20**).

Once a shoot achieves a diameter of 15–20 cm, its leafy tanks can retain moisture through much of the average year. At some point, a diverse assemblage of litter-feeding invertebrates and decomposer microbes assembles to assist the plant as it accumulates those substantial stocks of nutrients it needs to fuel its life-ending spasm of reproduction. Reaching this threshold is slow, especially until juveniles develop significant impounding capacity. Overall, more than a decade, and usually closer to two, pass before flowering occurs. What cues reproduction isn't known, but either the stimulus or the plant's response vary because reproducing shoots range severalfold in size.

Fruit set is accomplished with an erect, massive, bi- and even tri-pinnately branched inflorescence bearing 10 to 200 flowers on 5 to 40 branches that insert above a 20–50 cm long scape clothed with numerous overlapping bracts. The branches spread upward and outward, each one bearing laxly appressed bracts and even more laxly 6 to 11 flowered terminal portions that end in acute tips. Erect, well-separated, naked floral bracts 1.2–2.0 cm long leave much of the rachis exposed (**CP 14**). Color varies in intensity and location, and includes light pink-purple floral bracts with the same pigments sometimes highlighting much of the inflorescence axis. Shade retards pigment synthesis, and the entire reproductive apparatus, except for its petals, can be green.

Flowers open during the afternoon, and they may remain so into the night. The sepals are elliptic, 1.4–2.0 cm long, obtuse, and leathery; the corolla consists of white erect petals that twist modestly to form a slightly asymmetrically shaped tube up to 4 cm long. Both the stamens and stigma extend beyond the corolla. High fruit set indicates frequent self-pollination, although the flowers, being large and showy, probably encourage frequent visitations by animals, most likely moths after dark. Specimens representing populations beyond Florida often develop more color, but fewer of their flowers yield capsules.

Brownish cylindrical capsules up to 4 cm long release up

to 300 seeds rendered more airworthy by an umbrella-shaped, hairy appendage attached at one end. The seeds of *Tillandsia utriculata* are considerably heavier than those produced by *T. paucifolia,* and may be so compared to the other tillandsias that also propagate by offshoots in addition to seeds.

The single subspecies of *Tillandsia utriculata,* which ranges through much of peninsular Florida, occurs in two named forms. For our purposes, subspecies equals variety, the taxon designation used to describe similarly differentiated segregates

The following two are named segregates of *Tillandsia utriculata*:

of *Tillandsia fasciculata.*

TILLANDSIA UTRICULATA **SUBSP.** ***UTRICULATA*** **FORMA** ***UTRICULATA*** LINNAEUS

Synonym: *Tillandsia wilsoni* S. Watson, Pro. Amer. Acad. 23: 266. 1888.

This unusually large version of its species features gray-green foliage. Oaks are especially hospitable, and hundreds of plants may infest a single tree, often individuals that constitute the progeny of one particularly prolific parent. The exotic weevil *Metamasius callizona* is especially fond of this bromeliad, having already decimated its populations throughout much of central and south Florida. The name *Tillandsia lingulata* has been misapplied to this taxon.

TILLANDSIA UTRICULATA **SUBSP.** ***UTRICULATA*** **FORMA** ***VARIEGATA*** H. LUTHER

J. Bromeliad Soc. 28: 165–167. 1978.

Prominent longitudinal strips of largely chlorophyll-free, creamy white tissue alone distinguish this plant from its uniformly green (concolorous) relative. So far forma *variegata* is only known from Manatee County. See the description for *Guzmania monostachia* for additional discussion of leaf variegation among the bromeliads.

TILLANDSIA VARIABILIS

SCHLECHTENDAL

(CPs 35, 36)

Linnaea 18: 418. 1844.

Synonyms: *Tillandsia valenzuelana* A. Richard in Sagra Hist. Cuba 11: 267. 1850.

Tillandsia bouzeavii Chapman, Fl. S. U.S. ed 2, suppl.: 655. 1883.

Tillandsia variabilis qualifies as a medium-sized bromeliad among Florida's natives. Flowering shoots can be as much as 40 cm tall and possess 15 to 30, spirally arranged, 12–30 cm long leaves with ovate, pale to brownish bases capable of impounding modest amounts of water and debris. The narrowly triangular, lax blades break easily.

The simple to sparingly branched, upright, 3–10 cm tall inflorescence (**CP 36**) bears erect, overlapping, leaflike scape bracts. Five to 30 day-open, bird-adapted flowers distributed among the 1 to 3 flattened branches emerge from between overlapping, acute, scale-free, green to purple to red floral bracts. The lavender-blue, approximately 3 cm long tubular corolla comprised of strap-shaped petals extends above the naked, 1.5–1.8 cm long, ovate sepals. Stamens and stigma protrude beyond the corolla.

Flowering during spring through fall results in up to 3 cm long, three-parted, cylindrical capsules that release seeds equipped at one end with hairy, umbrella-shaped appendages. The name *variabilis* refers to the inconsistent manner in which the inflorescence of this bromeliad branches.

Habitats tend to be humid and shady—usually swamp forests dominated by hosts such as cypress, Carolina ash, red maple, and pond apple. This is a prolific species, sometimes densely colonizing low-hanging branches. Exposed to strong light in forest openings, its foliage takes on an attractive rosy hew.

Several of Florida's bromeliads interbreed often enough to maintain populations of hybrids, two of which bear Latin names: *Tillandsia x floridana* and *Tillandsia x smalliana.*

TILLANDSIA X FLORIDANA

L.B. SMITH

(CP 37)

Phytologia 57: 175–76. 1985.

Synonym: *Tillandsia fasciculata* var. *floridana* L.B. Smith, Phytologia 15: 197. 1967.

This putative interspecific hybrid flowers when up to 60 cm tall. Each of the shoots, which for well-established plants occur in small clusters, consists of 20 to 40 spirally arranged, gray, erect, 20–50 cm long leaves covered with coarsely appressed scales; the sheath is triangular, flat, and dark brown toward its base. A channeled, linear-triangular, leathery blade with an attenuated tip directs precipitation into a modest-sized impoundment below.

The bipinnately divided inflorescence with 3 to 20 semi-erect branches is itself upright, 15–30 cm tall, and densely covered with overlapping, leaflike bracts along its lower portion. Each of the 2–10 cm long, erect, flattened and acutely tipped, lateral branches bears overlapping elliptic, 2.0–2.5 cm long, erect, densely lepidote, pink floral bracts. Flowers that open during the day bear lanceolate, thin, leathery sepals 1.7–2.2 cm long located below a tubular corolla comprised of violet, strap-shaped petals up to 4.5 cm in length. The stamens and the stigma extend beyond the corolla tube. Cylindrical capsules up to 4 cm long release what appear to be viable seeds equipped with a hairy appendage at one end.

Flowering occurs from spring into summer. Plants grow epiphytically in swamps and river forests. *Tillandsia bartramii* and *Tillandsia fasciculata* var. *densipica* appear to be the parents. *Tillandsia x floridana* is easily confused with both *T. bartramii* and *T. simulata,* all three sharing about the same geographic distribution and similarly narrow foliage. Both the inflorescences and shoots of the two species are smaller. Specimens of *Tillandsia x floridana* have also been misidentified as *Tillandsia juncea,* a species with an exceptionally broad range, but not one that includes Florida.

TILLANDSIA X SMALLIANA

H. LUTHER

(CP 38)

Phytologia 57: 176. 1985.

Tillandsia x smalliana appears to be a natural hybrid between *T. balbisiana* and *T. fasciculata* var. *densispica*. Shoots flower when up to 50 cm tall and equipped with 20 to 40 spirally arranged, gray leaves covered with ash-colored scales. Small amounts of water accumulate during the wet season in the broadly elliptic, somewhat inflated, darkly rust-colored leaf sheaths that form a moderately swollen shoot base. From this bulblike foundation extend narrowly triangular, channeled to inrolled blades with much attenuated tips.

The inflorescence is upright, 15–35 cm tall and covered below with densely overlapping, erect to spreading scape bracts that diminish in size upward. Bipinnate branching yields 3 to 13 narrowly elliptic, flattened, laterals, each 2–6 cm long and acutely tipped. Elliptic, red, naked floral bracts 2.0–2.5 cm long overlap to obscure the rachis. The 5 to 40 day-open flowers feature lanceolate, 2.0–2.5 cm long, thin, nearly naked sepals topped by a tubular, violet corolla up to 5 cm long through which the stamens and stigma extend several more millimeters. The brownish, up to 3.8 cm long, three-part, cylindrical capsules contain what appear to be viable seeds equipped with a hairy umbrella-shaped appendage at one end.

Tillandsia x smalliana flowers from winter through summer. Although not common, its grows on diverse kinds of hosts, most often on Carolina ash, buttonwood, cypress, and oak, in well-lit forests and hammocks. It usually co-occurs with both of its putative parents.

Tillandsia x smalliana has been misidentified in some publications as *T. polystachia,* a structurally similar species distributed through much of the Caribbean, but nowhere in Florida. Hybrids attributable to the same parents reportedly also occur in Panama.

Neither *Tillandsia x floridana* nor *Tillandsia x smalliana* has received sufficient scrutiny to confirm its status as a hybrid. Should either one qualify, it would be worth trying to identify the responsible mechanisms. At first glance, *Tillandsia x smalliana* appears to possess structure about intermediate between that of the putative parents. Perhaps the entire population consists of first generation (F_1) progeny unable to backcross with either parent. Preliminary study has confirmed substantial pollen abortion for *T. x smalliana*—evidence, although not proof, of F_1 status. Should backcrossing to either parent be under way, proper analysis of the dimensions of leaves and certain parts of the reproductive apparatus will probably confirm it (as would DNA analysis), and also reveal how much and in which direction(s) gene flow has occurred.

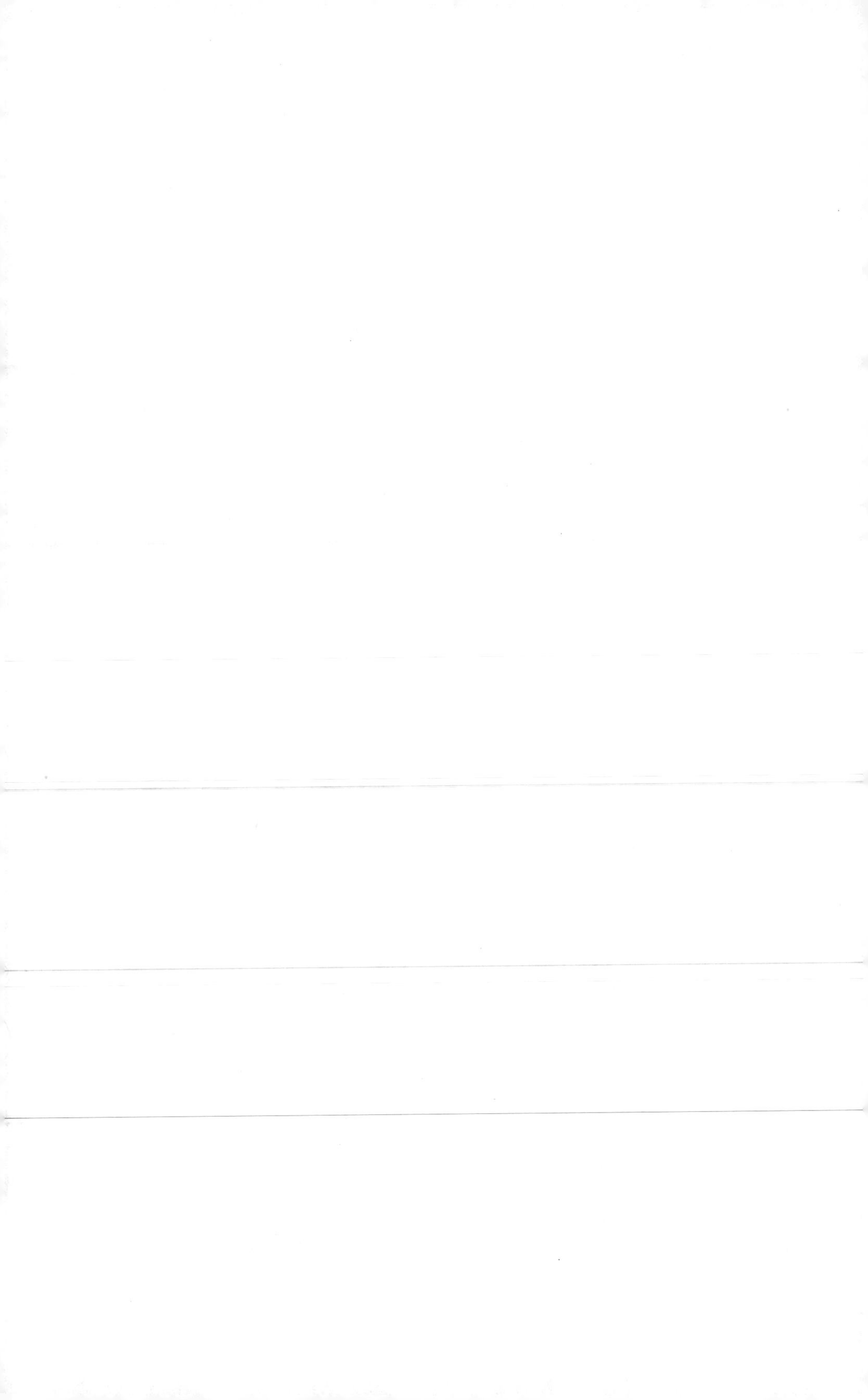

Tillandsia utriculata (top), *T. fasciculata* (bottom), Big Cypress National Preserve
(B. Holst photo)

CITED LITERATURE AND SUGGESTED READINGS

Benzing, D. H. 2000. *Bromeliaceae: Profile of an adaptive radiation.* Cambridge, UK: Cambridge University Press. 690 pp.

Benzing, D. H. 1990. *Vascular epiphytes*. Cambridge, UK: Cambridge University Press. 354 pp.

Benzing, D. H. 1981. The population dynamics of *Tillandsia circinnata* (Bromeliaceae): cypress crown colonies in Southern Florida. *Selbyana* 5: 248–55.

Crayn, D. M., K. Winter, and J. A. C. Smith. 2004. Multiple origins of crassulacean acid metabolism and the epiphytic habit in the Neotropical family Bromeliaceae. *Proceedings of the National Academy of Sciences* 101: 3703–8.

Frank, H. 1999–2000 (Winter). Florida's native bromeliads imperiled by exotic evil weevil. *The Palmetto* 19(4): 6–9.

Frank, J. H. and G. F. O'Meara. 1984. The bromeliad *Catopsis berteroniana* traps terrestrial arthropods but harbors *Wyeomyia* larvae (Diptera: Culicidae). *Florida Entomologist.* 67: 418–24.

Givnish, T., K. C. Millam, P. E. Berry, and K. J. Sytsma. 2007. Phylogeny, adaptive radiation, and historical biogeography of Bromeliaceae inferred from *ndh*F sequence data. *Aliso* 23: 3–26.

Laessle, A. M. 1961. A microlimnological study of Jamaican bromeliads. *Ecology* 42: 499–517.

Luther, H. E. 1993. A new record for *Tillandsia* (Bromeliaceae) in Florida. *Rhodora* 95: 342–7.

Martinez, J. D., M. Nathany, and V. Dharmarajan. 1971. Spanish moss, a sensor for lead. *Nature* 233: 564–65.

Pemberton, R. and H. Liu. 2007. Rare naturalization of an ornamental *Tillandsia, Tillandsia ionantha* in Southern Florida. *Selbyana* 28: 150–53.

Plever, H. and J. Brehm, editors. 2003. *Bromeliads: a Cultural Manual*. Bromeliad Society International, Inc.

Smith, L. B. and R. J. Downs. 1974. *Pitcairnioideae (Bromeliaceae).* Flora Neotropica Monograph No. 14, Part 1. New York: Hafner Press.

Smith, L. B. and R. J. Downs. 1977. *Tillandsioideae (Bromeliaceae).* Flora Neotropica Monograph No. 14, Part 2. New York: Hafner Press.

Smith, L. B. and R. J. Downs. 1979. *Bromelioideae (Bromeliaceae).* Flora Neotropica Monograph No. 14, Part 3. Bronx, NY: The New York Botanical Garden.

Ward, D. B. 1979. *Rare and Endangered Biota of Florida*. Vol. 5: *Plants*. Gainesville: University Presses of Florida.

GLOSSARY

Air plant: a term applied to certain epiphytes, especially the densely trichome-covered bromeliads like Spanish moss, that appear to be subsisting entirely on nutrients and moisture derived from the atmosphere

Anthesis: the process of flowering

Apical meristem: a group of well-organized embryonic stem cells located at the tips of roots and stems and responsible for their elongation

Appressed: closely and flatly pressed against something

Arboreal: describes organisms that live in the crowns of trees

Atmospherics: epiphytes that subsist primarily on nutrients that they extract from precipitation and other canopy fluids

Axil: the location above the insertion point where a leaf attaches to a stem, an axillary or lateral bud usually being present here

Backcross: a term that describes the process whereby an offspring resulting from a cross between two plants that represent two species is in turn interbred with an individual representing one of those parent species.

Bipinnate: an adjective that describes a pattern of branching whereby an organ such as a leaf is subdivided in two orders of parts

Biogeochemical cycle: the movements of mineral nutrients like phosphorous within ecosystems as the organisms that live there gain, lose, and exchange such substances between themselves and the physical components (e.g., the soil) of those systems

Bract: a leaf homolog, often small and scalelike usually protects something, like a developing flower in the case of a floral bract

Breeding system: a genetically based mechanism that determines paternity via sexual compatibility (e.g., self-incompatibility versus self-compatibility). *Also see* Dioecious

Bromeliaceae: the plant family to which the bromeliads belong

Bromelioideae: one of the three subfamilies of Bromeliaceae

C_3: a designation for the more primitive of the two mechanisms of photosynthesis that occur among the bromeliads

CAM: (crassulacean acid metabolism) the photosynthetic mechanism that allows plants to take up carbon dioxide at night to enhance water-use efficiency

Capsule: a type of fruit that contains numerous seeds and is dry and splits open at maturity

Carnivore: an organism that subsists on captured prey for nutrition

Caulescent: long-stemmed, the opposite of rosulate in describing the structure of a shoot or ramet

Coriaceous: relatively thick and leathery as applied to foliage

Concolorous: of uniform color, usually green on both sides, as in leaves. *See* Discolorous

Corolla: all of the petals of a flower

Deacidification: the step in the CAM mechanism that involves the break-down of stored malic acid during the day to regenerate carbon dioxide for making sugar using light

Decomposers: microorganisms that feed on dead biomass reducing it to its basic mineral constituents

Determinate: an adjective that describes a plant or certain of its parts that are genetically programmed to grow to a predetermined size or produce only a predetermined number of parts, e.g., the individual leaf and the ramets of bromeliads

Dioecious: describes a species that consists of individuals that produce flowers with either functional male or female appendages, but not both

Discolorous: of different shades of green, or of two different colors (said of leaves)

Distichous: an adjective that describes shoots that bear leaves in two opposite, alternating ranks

Diurnal: opening, as for certain kinds of flowers, or occurring during the day; opposite of nocturnal

Ecosystem: a physical space and all of the organisms and nonliving entities like soil located therein

Endemic: describes an organism when growing within its native range

Epiphyte: a plant that grows on another plant, **facultative** (i.e., sometimes) or **obligate** (i.e., always)

Exotic, alien: describes an organism or population when it is growing beyond its native habitat, i.e., where it is not endemic

Extirpation: extinction

Filiform: long and narrow, as referring to the shape of the foliage of many bromeliads

Free-living: for epiphytes meaning not parasitic like the mistletoes

Haustorium: the invasive organ of a parasitic plant, e.g., a mistletoe

Hemi-epiphyte: a plant that regularly spends part of its life rooted in the ground and the rest anchored on a host

Herbivore: an animal that feeds primarily on plant material

Imperfect flower: a flower that contains either functional male or female appendages (stamens or pistils), but not both

Inflorescence: that part of a plant dedicated to the production of flowers and fruits

Interspecific hybrid: an offspring that results when members of two different species interbreed

Invasive species: species that invade territory and spread aggressively, often displacing other species, often non-natives

Involute: describes a flat body like a bract that is curved inward or upward

Lanceolate: lance-shaped

Lepidote: covered with small scales or trichomes

Lifestyle: a way of living, including how resources are obtained and from where

Lithophyte: a plant that normally grows on rocks

Litter: plant parts that have been shed after their functional life is finished, e.g., spent foliage

Malic acid: an organic acid abundant in many edible fruits, also synthesized at night by CAM plants as a way to capture and store carbon dioxide for later synthesis into sugar

Mistletoe: a parasitic epiphyte

Module: a term used to describe the serially produced structural subunits that make up the bodies of many herbaceous perennials, equivalent to a ramet when applied to bromeliads

Monocarpic: describes plants that flower just once and then die

Mutualist: an organism that lives in close association with a partner representing another species. Both parties benefit from their association with the other, i.e., they engage in a mutualism

Myrmecodomatium: a structure provided by some plants to entice ants to nest inside their bodies

Naturalized: describes a species or population where it has become a permanent resident, i.e., is reproducing, beyond its native habitat

Nectary: a gland that secretes nectar, either floral or extrafloral nectar depending on its location on the plant

Nerve: a vein in a leaf or bract

Nutritional piracy: a term used to define the process whereby free-living epiphytes obtain nutrients from their hosts by capturing them as they cycle between the host's crown and roots as part of the biogeochemical cycling process

Phytotelm: (plural phytotelms) [also spelled phytotelma (plural phytotelmata)]: a plant-provided cavity, for example, the leafy tank of a bromeliad, which impounds water and is often inhabited by aquatic organisms

Pinnate: describes an organ that is subdivided once into parts that align like the parts of a feather, on either side of a central axis, as applied to compounded leaves and branched inflorescences

Pistil: the female appendage of a flower consisting of a basal ovary topped by a style and stigma

Pitcairnioideae: one of the three subfamilies that comprise family Bromeliaceae

Polycarp: describes a perennial plant that fruits repeatedly over successive seasons, the opposite of a monocarp

Polymorphism: describes a population or species that exhibits much genetically based variation among its members

Propagule: a dispersed reproductive organ of a plant whether a seed, spore, or a vegetative structure

Proto-carnivore: a plant equipped with only rudimentary adaptations for prey use

Pup: colloquial designation for an immature ramet

Rachis: the axis upon which the ultimate parts of divided appendages are attached, e.g., flowers, leaflets of compound leaves

Ramet: a determinate shoot that represents a module

Rosulate: describes a shoot that consists of a short stem bearing congested spirally attached leaves

Scape: an axis of an inflorescence that lacks bracts in the region located below the terminal part that bears flowers

Self-compatible: describes plants that can set seeds with their own pollen, a type of breeding system

Self-incompatible: describes plants that cannot set seeds with their own pollen, a type of breeding system

Sepals: the outermost set of appendages of a flower

Sessile: describes an organ that is directly attached to its supporting plant part, e.g., flowers on an inflorescence

Shoot: that part of the plant body made up of stems, leaves, and flowers and sexual reproductive organs

Spike: a type of inflorescence characterized by an unbranched axis that bears multiple flowers that lack pedicles, i.e., are sessile

Stamen: the male appendage of a flower, comprised of a filament and an anther, the pollen-bearing portion

Stigma: the pollen-receptive apical portion of a pistil

Stomata: tiny organs located in the epidermis of leaves and other parts of the shoot. It consists of a central pore (stoma) that opens and closes as the shapes of the adjacent pair of guard cells change; stomata allow the plant to control the uptake of carbon dioxide and loss of water (gas exchange)

Succulent: describes a leaf or stem that is much thickened by the presence of water storage tissue

Tank: a leafy impoundment in which litter and water accumulate, characteristic of many bromeliads and some other epiphytes. *Also see* Phytotelm

Taxon (singular), **Taxa** (plural)**:** a term used to designate a taxonomic category in the Linnaean hierarchy used to classify organisms, e.g., species, genus, and family

Terrestrial: describes a plant that roots in the ground

Tillandsioideae: one of the three subfamilies that comprise family Bromeliaceae

Transpiration: the process of losing water vapor through open stomata

Transpiration ratio: the weight-weight ratio that expresses the amount of water expended to produce biomass through the photosynthetic process; a measure of water use efficiency

Trichome, scale, hair: terms applied to the characteristic epidermal appendage that almost all of the bromeliads bear on their foliage, which for the more advanced species allows them to dispense with absorbing roots

Velamen: the multilayered epidermis that covers the roots of many orchids and assists function in air

INDEX

Note: Figures and illustrations are indicated by boldface type.
CP refers to the color plates between pages 64 and 65.

ABOUT THE AUTHORS

Harry Luther, born in St. Petersburg, Florida, is the Director of the Mulford B. Foster Bromeliad Identifica-tion Center and Curator of the Living Collections at the Marie Selby Botanical Gardens in Sarasota, Florida. With forty years of fieldwork in bromeliad habitats, including Florida, he is a much-requested speaker for horticulture and botanical groups nationally and internationally, and has published over a hundred papers both technical and popular.

David Benzing received his Ph.D. in botany from the University of Michigan and is retired from teaching biology at Oberlin College. His publications, which include four books, chapters in about twenty more volumes, and approximately seventy-five reports in technical journals, mostly deal with the adaptive biology of vascular epiphytes, especially bromeliads and orchids. He currently holds the Jessie B. Cox Chair for Tropical Botany at the Marie Selby Botanical Gardens in Sarasota, Florida, where he spends winters. Summers are spent in northern Ohio, where he and two partners own and manage a small vineyard and winery that operate according to ecologically sustainable land use and energy practices.

Here are some other books from Pineapple Press on related topics. For a complete catalog, visit our website at www.pineapplepress.com. Or write to Pineapple Press, P.O. Box 3889, Sarasota, Florida 34230-3889, or call (800) 746-3275.

Tropical Trees of Florida and the Virgin Islands: A Guide to Identification, Characteristics and Uses by T. Kent Kirk. This user-friendly, all-color field guide aids in the identification of more than 90 species of trees native to Florida and the Virgin Islands (and a few widespread exotics). For each species, there are photos of the whole tree, leaves, flowers, and fruit. (pb)

Trees of Florida: A Reference and Field Guide by Gil Nelson. The first comprehensive guide to Florida's amazing variety of trees, both natives and exotics, from scrub oak to mangroves, from bald cypress to gumbo limo, from sabal palm to the Florida yew. Includes suggested field sites for observing the species described. With color photos and line drawings. (hb & pb)

The Ferns of Florida: A Reference and Field Guide by Gil Nelson. A complete guide to Florida's amazing variety of ferns. Color plates feature more than 200 images, some of which include rare species never before illustrated in color. (hb & pb)

The Shrubs & Woody Vines of Florida: A Reference and Field Guide by Gil Nelson. This easy-to-use field guide includes more than 550 woody vines and shrubs native to Florida. With color photos and line drawings. (hb & pb)

Natural Florida Landscaping by Dan Walton and Laurel Schiller. This book will help you make a plan that will work for your yard and choose the native plants that will thrive there in order to create a beautiful and environmentally sensitive landscape. (pb)

Florida's Best Fruiting Plants by Charles R. Boning. A comprehensive guide to fruit-bearing plants that thrive in the Florida environment. Discusses exotics and native species, familiar plants, and dozens of rare and obscure plants. (pb)

100 Orchids for Florida by Jack Kramer. One hundred beautiful orchids you can grow in Florida, chosen for their beauty, ease of cultivation, and suitability to Florida's climate. (pb)

Flowering Shrubs and Small Trees for the South by Marie Harrison. Author and Master Gardener Marie Harrison offers tips on how to identify, select, and care for more than 100 flowering shrubs and small trees suited to the South. Full-color photos and line drawings throughout. (pb)

Groundcovers for the South by Marie Harrison. Each entry gives detailed information on ideal growing conditions, plant care, and different selections within each species. Color photographs and line drawings make identification easy. (pb)

Southern Gardening: An Environmentally Sensitive Approach by Marie Harrison. A comprehensive guide to beautiful, environmentally conscious yards and gardens. Suggests useful groundcovers and easy-care, adaptable trees, shrubs, perennials, and annuals. (pb)

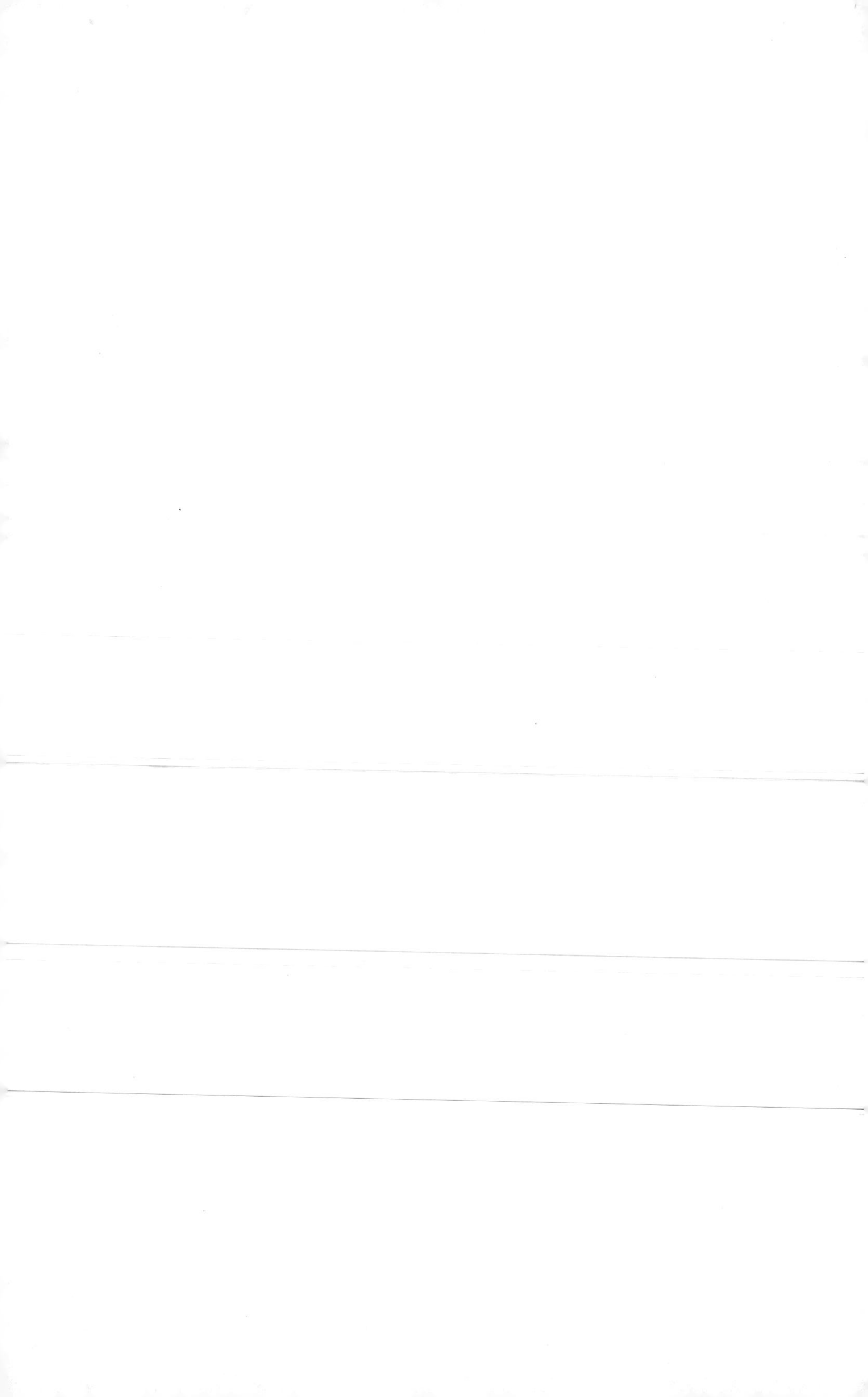